Understanding Basic Statistics

SIXTH EDITION

Charles Henry Brase
Regis University

Corrinne Pellillo Brase
Arapahoe Community College

Prepared by

Linda Myers
HACC, Central Pennsylvania's Community College

Australia • Brazil • Japan • Korea • Mexico • Singapore • Spain • United Kingdom • United States

ISBN-13: 978-1-133-52509-7
ISBN-10: 1-133-52509-1

Brooks/Cole
20 Channel Center Street
Boston, MA 02210
USA

Cengage Learning is a leading provider of customized learning solutions with office locations around the globe, including Singapore, the United Kingdom, Australia, Mexico, Brazil, and Japan. Locate your local office at: **www.cengage.com/global**

Cengage Learning products are represented in Canada by Nelson Education, Ltd.

To learn more about Brooks/Cole, visit **www.cengage.com/brookscole**

Purchase any of our products at your local college store or at our preferred online store **www.cengagebrain.com**

Printed in the United States of America
1 2 3 4 5 6 7 16 15 14 13 12

Table of Contents

CHAPTER 8: ESTIMATION

CHAPTER 9: HYPOTHESIS TESTING

CHAPTER 10: INFERENCES ABOUT DIFFERENCES

CHAPTER 11: ADDITIONAL TOPICS USING INFERENCE

TO THE STUDENT

This Notetaking Guide can be used with *Understanding Basic Statistics*, Sixth Edition. It is a supplement to the textbook by Charles and Corrinne Brase. All references to chapters, sections, definitions, and examples apply to this textbook.

How to Use the Notetaking Guide

Each section of the Notetaking Guide corresponds to a textbook section, allowing you to study the sections assigned by your teacher. The Notetaking Guide is designed to be flexible, so you can use the material in the way that best fits your style of studying. See below for some suggestions on using the Notetaking Guide to prepare for class, take notes, study, and review for tests.

Using the Notetaking Guide to prepare for class:
- In the textbook, read the section you will cover in class.
- Work through each part of the Notetaking Guide for that section, filling in blanks and working each example.
- Make a note of questions you have about a topic so you can ask about it later.

Using the Notetaking Guide during class:
- Bring the section you will cover that day to class with you.
- As the teacher covers each part of the material, take notes directly in the Notetaking Guide about vocabulary, concepts, and examples. Use this to supplement the notes you take in your regular notebook.
- When the teacher finishes covering a topic, check your notes to make sure all your questions about that topic are answered. Ask questions about any concept you do not understand before the teacher moves on to the next topic.

Using the Notetaking Guide to review and prepare for tests.
- After each class, review your notes in your notebook and the Notetaking Guide and fill in any gaps. If you still have questions, make a note of them to ask your teacher.
- Read your notes in your textbook and in the Notetaking Guide before tests to make sure you understand key vocabulary, concepts, and formulas.
- If you need further practice working on a topic, check the Homework Assignment for that section to see which exercises you have already finished. This will allow you to work through review exercises for that section that you have not yet tried.

Chapter 1 Getting Started

Section 1.1 What is Statistics?

<table>
<tr><td>Course Number</td></tr>
<tr><td>Instructor</td></tr>
<tr><td>Date</td></tr>
</table>

Objective:
In this section you learned how to identify variables in a statistical study, how to distinguish between quantitative and qualitative variables, how to identify populations and samples, how to distinguish between parameters and statistics, how to determine level of measurement, and how to compare descriptive and inferential statistics.

Important Vocabulary   Define each term or concept.

Statistics

Individuals

Variable

Quantitative Variable

Qualitative Variable

Population Data

Sample Data

Parameter

Statistic

Nominal Level of Measurement

Ordinal Level of Measurement

Interval Level of Measurement

Ratio Level of Measurement

Descriptive Statistics

Inferential Statistics

Brase/Brase *Understanding Basic Statistics, Sixth* Edition, Student Notetaking Guide   **1**

I. Introduction

A characteristic of the _______________ to be measured or observed in a __________ study is called a _______________.

The variables such as body weight or employee's salary are _____________ variables. Variables such as gender or nationality are _____________ variables.

The entire group of individuals of interest is called the _______.
A subgroup of that is called a _____________.

A numerical measure that describes an aspect of the entire population is a __________, while a numerical measure obtained from a sample is called a __________.

Example 1. A college with a total enrollment of 7000 students wants to know what proportion of its students use the library at least 10 hours a week. The college asked 100 students whether they spend at least 10 hours per week in the library or not.

The individuals of the study are _____________________ and the variable is _____________.

Do the data comprise a sample? If so, what is the underlying population?

Is the data quantitative or qualitative?

Is the proportion of students in the sample who use the library at least 10 hours a week a statistic or a parameter?

II. Level of Measurement: Nominal, Ordinal, Interval, Ratio

At the _____________ level of measurement, we can put the data into categories; at the ___________ level, we can order the data from smallest to largest; at the ___________ level, we can order the data and also take the differences between data values; at ________ level, we can order the data, take differences, and also find the ratio between data values.

Example 2. The following are some data about a college professor. For each data entry, indicate the corresponding level of measurement.

Focus Points
how to recognize variables in a statistical study, quantitative and qualitative variables, populations and samples, parameters and statistics

Focus Points
how to determine level of measurement

Brase/Brase *Understanding Basic Statistics,* Sixth Edition, Student Notetaking Guide

1. The professor's name is James Smith.
2. The professor is 50 years old.
3. The professor has been working at the college for three years: 2000, 2001, 2002.
4. The professor is ranked 17th in his department's seniority list.

III. Looking Ahead

Organizing, picturing, and summarizing information from all individuals in a sample or a population are the methods of __________ statistics; while drawing conclusions regarding the population based on the study of a sample is called the ______________ statistics.

> **Focus Points**
> how to compare descriptive and inferential statistics.

Additional Notes

Homework Assignment

Page(s)

Exercises

Section 1.2 Random Samples

Objective:
In this section you learned how to explain the importance of random samples, how to construct a simple random sample using random numbers, how to simulate a random process, and how to describe stratified sampling, cluster sampling, systematic sampling, multistage sampling, and convenience sampling.

<table>
<tr><td>Course Number</td></tr>
<tr><td>Instructor</td></tr>
<tr><td>Date</td></tr>
</table>

Important Vocabulary   Define each term or concept.

Simple random sample

Random number table

Simulation

Sampling with replacement

Stratified sampling

Systematic sampling

Cluster sampling

Multistage sampling

Convenience sampling

Sampling frame

Undercoverage

Sampling error

I. Simple Random Samples

A sample in which every individual in the population has an equal chance to be selected is called a

________________________________.

Using a group of pre-generated random numbers, you can form a ____________________.

Example 1. To play a game, you pay one dollar and choose three different numbers from the group of numbers 1 through 10. If your group of three numbers matches the winning group of three numbers selected by simple random sampling, then you win one hundred dollars.

    (a) Is the number 4 as likely to be selected in the winning group of three numbers as the number 8?

    (b) Could all the winning numbers be odd?

    (c) If you always play the numbers 2, 4, 6, could you ever win?

Example 2. Use a random-number table to pick a random sample of 10 students from a class of 50 students.

II. Simulation

Using numbers to represent real-world objects is the procedure of ____________. If the same number may be selected for a sample more than once, then this sampling is called ________________

____________.

III. Other Sampling Techniques

Dividing the population into distinct subgroups and then drawing random samples from each subgroup is called ____________.

Sequentially numbering all the individuals in the population, and then from a randomly selected starting point, taking every k^{th} member for the sample is called ____________.

Dividing the population into clusters and then making a random selection of clusters is called ____________.

Creating smaller groups stage by stage using various sampling methods, with the final sample consisting of clusters is called ____________.

> **Focus Points**
> how to construct a simple random sample and how to use random number table

> **Focus Points**
> how to conduct simulation and sampling with replacement

> **Focus Points**
> how to describe stratified, systematic, cluster, multistage, and convenience samples, sampling frame, undercoverage, and sampling error

Using available data from the population to make a sample is called

_______________________________.

The list of members from which a sample is actually selected is called
the _____________________. Since part of the population is omitted in
the selection, _________________ occurs.

The difference between measurements from a sample and corresponding
measurements from the respective population is called the ____________
_________.

Additional Notes

Homework Assignment

Page(s)

Exercises

Section 1.3 Introduction to Experimental Design

Objective:
In this section you learned what it means to take a census, how to describe simulations, observational studies, and experiments, how to identify control groups, placebo effects, completely randomized experiments, and randomized block experiments. You also saw potential pitfalls that might make your data unreliable.

Course Number

Instructor

Date

Important Vocabulary	Define each term or concept.

Census

Sample

Observational study

Experiment

Placebo effect

Completely randomized experiment

Block

Randomized block experiment

Control group

Randomization

Replication

Nonresponse

Lurking variable

Confounding variable

Important Vocabulary   Define each term or concept.

Hidden bias

Vague wording

Interviewer influence

Voluntary response

Lurking variable

Confounding variable

I. Planning a Statistical Study

Observations from the entire population are used in a ______,
while observations from part of the population are used in a
__________.

Focus Points
how to describe the
difference between a
census and a sample

II.   Experiments and Observation

An _________________ is conducted in a way that doesn't change
the response or the variable being measured. In an __________, a
treatment is used to inflict a possible change in the variable being
measured.

Using numbers to represent real-world objects is the procedure of
_____________.

Example 1. An elementary school teacher wanted to see how extra
exercises would improve students' grades. She constantly assigned
extra exercise problems to one of her classes, but for her other
class she just gave the regular assignments. At the end of year she
compared the two classes. Is this an observational study or an
experiment?

Focus Points
how to describe
simulations,
observational
studies, and
experiments;
how to identify
control groups,
placebo effects,
completely
randomized
experiments,
and randomized
block
experiments

When a random process is used to assign each member
to one of the treatments, it is a ___________________.
If someone thinks that he or she is receiving treatment and
responds favorably but in fact no treatment is given to this
person, it is called a _________.

A group of members having certain common features that
might affect the treatment forms a ___________. A process
of first forming these groups and then randomly assigning
each member in the group to one of the treatments is called a

___________________.

To account for the influence of other known or unknown
variables that might cause changes in response, we use a

___________________.

To assign members to the two treatment groups in order to
prevent bias in selecting members for each group, we use

___________________.

To reduce the possibility that the differences between the two
groups occurred by chance alone, we _______________ the
experiment.

Example 2. If the elementary school teacher in Example 1 teaches
only one class, she could randomly assign her students into two
groups and give extra exercise problems to one group but not the
other. Then at the end of the year she can compare the performance
of the two groups. This is one example of a completely randomized
experiment.

Example 3. Which technique of gathering data (sampling, experiment,
simulation, or census) do you think might be the most appropriate for
the following studies?

   (a) Effect of a large scale earthquake.
   (b) Amount of time high school students spend online each week.

   (c) Effect of a new medicine for a certain heart disease.

   (d) Whether wedding guests are allergic to the planned main course.

III. Surveys

Focus Points
how to recognize potential pitfalls of a survey, lurking variables, and confounding variables

Some potential pitfalls of a survey include: ___________ when no response is obtained from some individuals, ___________ when the respondents may not be telling the truth, ___________ when the respondents may not remember the truth, ___________ when the question may be worded in such a way as to elicit a certain response, ___________ when the words such as "frequently," "sometimes," and "quite good" mean different things to different people, _______ when the attitudes of the interviewer might influence responses, and _______________ when individuals with strong feelings about a subject are more likely than others to respond.

The study of the cause-and-effect relationships between two or more variables can be complicated by a _______________ for which no data have been collected but that nevertheless has influence on other variables in the study, or by _________________ when the effects of one variable cannot be distinguished from the effects of the other.

Example 4. What is the potential pitfall of the data collected as described?

(a) A school principal calls ten students to his office. He writes down each student's name and asks whether he or she has cheated on any tests in the past year.

(b) A survey about changing the graduation requirements was conducted by mailing survey forms to students. A drop box for completed forms was available by the main entrance of the school.

Homework Assignment

Page(s)

Exercises

     Brase/Brase *Understanding Basic Statistics,* Sixth Edition, Student Notetaking Guide

Chapter 2 Organizing Data

Section 2.1 Frequency Distributions, Histograms, and Related Topics

<table>
<tr><td>Course Number</td></tr>
<tr><td>Instructor</td></tr>
<tr><td>Date</td></tr>
</table>

Objective:
In this section you learned how to organize raw data using a frequency table, how to construct histograms and relative-frequency histograms, how to recognize basic distribution shapes, and how to interpret graphs in the context of the data setting.

Important Vocabulary   Define each term or concept.

Frequency table

Class lower limit

Class upper limit

Class width

Class frequency

Class midpoint

Upper class boundary

Lower class boundary

Relative frequency

Relative frequency table

Histogram

Relative-frequency histogram

Ogive

Important Vocabulary   Define each term or concept.

Mound-shaped symmetrical distribution

Uniform or rectangular distribution

Skewed left distribution

Skewed right distribution

Bimodal distribution

Outliers

I. Frequency tables

Data are divided into classes. The lowest data value in a class is called the ______________ and the highest value is the __________ ________. The average of them is called the ____________. The difference between the lower class limit of one class and the lower class limit of the next class is called the ____________. The number of values in a class is called the __________________. A table showing the classes and corresponding frequencies is called a ______________________. The ______________________ of a class equals its frequency divided by the sample size.

> *Focus Points*
> how to organize raw data using frequency table

The ________________ equals the upper limit plus 0.5 and the ______________ equals the lower limit minus 0.5.

Example 1. Given a data set of 10 numbers {1, 7, 8, 4, 4, 5, 6, 3, 8, 7}, construct a frequency table using four classes.
    (a) The class width = ________________.

(b) The lower class limits are _________________.
(c) The upper class limits are _________________.
(d) The class boundaries are _________________________.

(e) The class midpoints are _________________.
(f) Make a frequency table below.

(g) The relative frequencies for these four classes are _______________.

(h) Make a relative frequency table below.

II. Histograms and Relative Frequency Histograms

A bar chart where the width of the bar equals the class width, and the height of the bar equals the class frequency is called a __________. Sometimes we use the relative frequency as the height. Then we have a _________________________.

Focus Points
how to construct histograms and relative frequency histograms

Example 2. Make a histogram and a relative-frequency histogram for the data in Example 1.

III. Distribution Shapes

A histogram has a ___________________________ if its two sides are
symmetric with respect to the vertical line that goes through
the middle of the graph. A histogram has a ________________ if
every class has the same frequency. If the tail on the left is
longer than the one on the right, then the histogram is ________
____________. If the tail on the right is longer, then the histogram
is ______________________. If a histogram shows two "peaks",
that is, there are two classes with the largest frequencies that are
separated by at least one class, then the histogram has a __________

__________.

<table>
<tr><td>Focus Points
how to recognize basic
distribution shapes and
interpret graphs in the
context of the data setting</td></tr>
</table>

In some data sets, a few values are so high or so low that they are far away from the rest of
the data. These values are called ___________.

<table>
<tr><td>

Homework Assignment

Page(s)

Exercises

</td></tr>
</table>

Section 2.2 Bar Graphs, Circle Graphs, and Time-Series Graphs

Course Number

Instructor

Date

Objective:

In this section you learned how to determine types of graphs appropriate for specific data, how to construct bar graphs, Pareto charts, circle graphs, and time-series graphs, and how to interpret information displayed in graphs.

Important Vocabulary   Define each term or concept.

Bar graph

Clustered bar graph

Pareto chart

Circle graph (pie chart)

Time series graph

Time series data

For each of the following statements, determine whether the statement is true or false.

> *Focus Points*
> how to determine types of graphs appropriate for specific data, how to construct bar graphs, Pareto charts, circle graphs, and time-series graphs, and how to interpret information displayed in graphs.

(a) In a bar graph, the bars do not have to be of uniform width.

(b) The bars in a bar graph can be vertical or horizontal.

(c) The lengths of the bars in a bar graph stands for certain values of the variable being displayed.

(d) When two or more variables are displayed together, the bar graph is called a clustered bar graph.

(e) In a Pareto chart, the bars are arranged from left to right according to increasing height.

(f) A circle graph is also called a pie chart.

(g) Circle graphs are usually used to display percentages.

(h) A time series data contains the values of a variable taken at regular intervals over a certain time period.

(i) A time series graph displays data in the corresponding time series data.

Example 1. Below is the data (in hundreds) showing the enrollment of male students and female students in a college in the years of 1995, 2000, and 2005. Make a clustered bar graph for this data.

Year	1995	2000	2005
Male	30	34	32
Female	28	35	33

Example 2. Use the data given in Example 1, make a Pareto chart for male student enrollment in those three years.

Example 3. Use the data given in Example 1, make a circle graph to display the distribution between male and female students in 1995's enrollment.

    (a) Fill in the missing parts in the table below. Remember that the central angle of a circle is $360°$. Round to the nearest degree.

Students	Number (in hundreds)	Fractional Part	Percentage	Degrees
Male	30	30/58	___	___
Female	28	_____	___	___
Total	58	___	___	___

    (b) Make the circle graph.

Example 4. Use the data given in Example 1, make a time-series graph for the female student enrollment in those three years.

Additional Notes

Homework Assignment

Page(s)

Exercises

Brase/Brase *Understanding Basic Statistics,* Sixth Edition, Student Notetaking Guide

Section 2.3 Stem-and-Leaf Displays

Objective:
In this section you learned how to construct a stem-and-leaf display from raw data, how to use a stem-and-leaf display to visualize data distribution, and how to compare a stem-and-leaf display to a histogram.

Course Number

Instructor

Date

Important Vocabulary   Define each term or concept.

Exploratory data analysis

Stem-and-leaf display

Split stem

Back-to-back stem

I. Exploratory Data Analysis

Focus Points
how to describe exploratory data analysis

_____________________ techniques are used to explore a data set, to detect patterns and extreme data values, to raise new questions, or to pursue leads in many directions.

II. Stem-and-Leaf Display

Focus Points
how to make a stem-and-leaf display

A stem-and-leaf display is used to _________ and arrange data into groups. In a stem-and-leaf display, the _______ are aligned vertically from smallest to largest. A vertical line is drawn to the right of the stems. _________ with the same stem are placed in the same row as the stem, arranged in ___________ order. A label is used to indicate the magnitude of the numbers in the display.

Example 1. A study on peanut butter reported the following optimal consumption temperatures for various brands:

56 44 62 36 39 53 50 65 45 40

Brase/Brase *Understanding Basic Statistics,* Sixth Edition, Student Notetaking Guide   **19**

Make a stem-and-leaf display for this data.

Example 2. For the following data, use the first two digits as the stem to make a stem-and-leaf display.

106 94 112 96 89 113 90 85 85 100

Example 3. Looking at the distributions in Example 1 and Example 2, would you say that they are fairly symmetrical?

Homework Assignment

Page(s)

Exercises

Chapter 3 Averages and Variation

Section 3.1 Measures of Central Tendency: Mode, Median, and Mean

<table>
<tr><td>Course Number</td></tr>
<tr><td>Instructor</td></tr>
<tr><td>Date</td></tr>
</table>

Objective:
In this section you learned how to compute mean, median, and mode from raw data, how to interpret what mean, median, and mode tell you, how to explain how mean, median, and mode can be affected by extreme values, what a trimmed mean is and how to compute it, and how to compute a weighted average.

Important Vocabulary   Define each term or concept.

Mode

Median

Mean

Summation symbol, Σ

Resistant measure

Trimmed mean

Weighted average

I. Mode and Median

If a value in a data occurs most frequently in the data, then this value is called the __________ of the data.

> **Focus Points**
> how to compute median, and mode, how to interpret what median, and mode tell you

Order the values in a data from smallest to largest. If the data has an odd number of data values, then there will be only _____ middle value. This middle value is called the _______. If the data has an even number of data values, then there will be _____ middle values, and the average of them is the __________ of the data.

Example 1. Find the mode of this data: {1, 7, 8, 4, 4, 4, 6, 3, 8, 7}.

Example 2.

    (a) For the data in Example 1, find the median. Note that there are 10 (an even number) data values in this data.

    (b) If the smallest data value is dropped from the above data, then there will be on 9 (an odd number) data values in that data. What is the median now?

    (c) If the above data contains the amount of monthly earnings (in million dollars) of a company in the past 10 (or 9) months, and either the mode or the median should be used to describe the achievements of the company to the public, which value would the company most likely to use?

II. Mean and Trimmed Mean

The average of all values in a data is called the _____ of the data.

The summation symbol meaning "sum the following" is _____.

Example 3. To graduate, Steve must have at least a C in history. He did fairly well on the first four tests; however, he failed the last one. Here are his scores

73   80   69   72   35

Find the mean score and determine whether Steve will get at least a C (70 average or better).

Clearly, Steve can not obtain a C because his score on the last test is "extremely" low. A ____________________ is one that is not influenced by extremely high or low data values. One measure of this kind is the ____________ which is the mean of the data values left after "trimming" a specified percentage of the smallest and largest data values from the data set.

Example 4. Continue with Example 3. Suppose the school allows Steve to drop the highest and the *lowest* scores, that is, to take a 20% trimmed mean of his scores.

> **Focus Points**
> how to compute mean and how the mean can be affected by extreme values, what a trimmed mean is and how to compute it

(a) What are the remaining scores?

(b) What is the 20% trimmed mean (that is,
   what is the mean of remaining scores)?

(c) Would Steve get a C for the course if the 20%
   trimmed mean is used?

(d) What are the medians of the original score data
   and the trimmed score data? Does trimming change
   the value of median?

(e) Is the trimmed mean or the original mean closer
   to the median?

III. Weighted Average

Focus Points
how to compute a
weighted average

When weight (such as "importance") is assigned to values in a data,
the ________________ of the data can be computed by multiplying each
value by its __________ and adding the results together, and then diving
by the ____________________.

Example 5. Suppose your midterm test score is 80, your project score
is 85, and your final exam score is 98. Suppose the weights are 30% for
midterm, 30% for project, and 40% for final. If the minimum average of
an A is 90%, will you earn an A?

Additional Notes

Additional Notes

Homework Assignment

Page(s)

Exercises

Section 3.2 Measures of Variation

<table>
<tr><td>Course Number</td></tr>
<tr><td>Instructor</td></tr>
<tr><td>Date</td></tr>
</table>

Objective:
In this section you learned how to find the range, variance, and the standard deviation, how to compute the coefficient of variation from raw data, why the coefficient of variation is important, and how to apply Chebyshev's theorem and what a Chebyshev interval tells us.

Important Vocabulary   Define each term or concept.

Range

Sample variance

Sample standard deviation

Population mean

Population variance

Population standard deviation

Coefficient of variation

Chebyshev's theorem

Mean of grouped data

I. Range

The largest value minus the smallest value of a data gives the
__________ of the data.

Example 1. Given the following two samples

Sample 1: {10, 14, 15, 16, 20}

Sample 2: {10, 11, 15, 19, 20}

    (a) Find the range of each.

    (b) Find the mean of each.

    (c) Do the two samples have the same range and mean? How
        do the samples differ?

> *Focus Points*
> how to find the
> range

II. Variance and Standard Deviation

To compute the __________________, first take the sum of squares of
deviations, then divide that sum by $(n-1)$, where n is the sample size.
The square root of this result is called the ____________________
___________.

To compute the __________________, first take the sum of squares of
deviations, then divide that sum by the sample size n. The square root
of this result is called the ____________________________.

> *Focus Points*
> how to find the
> variance and the
> standard deviation

Example 2. For the two samples in Example 1, find the sample variance
and sample standard deviation.

Example 3. For the sample {1, 3, 2, 6}, follow the steps below to find
out the standard deviation s.

(a) Construct this table:

x	x^2
1	
3	___
2	___
6	___
$\Sigma x =$ ___	$\Sigma x^2 =$ ___

(b) What is the value of n? of $n-1$? Use the computation formula to find the sample variance s^2.

(c) What is the value of the sample standard deviation?

III. Coefficient of Variation

To express the standard deviation as a percentage of the sample or population mean, we use the ________________ , which is calculated by dividing the ________________ by the mean and multiplying the result by ____ .

> **Focus Points**
> how to compute the coefficient of variation and why the coefficient of variation is important

Example 4. For a *population* of five values $\{1, 4, 4, 3, 5\}$,

(a) Find the population mean μ and population standard deviation σ.

(b) Find the coefficient of variation CV.

Example 5. For the two samples given in Example 1,

(a) Find their sample mean and sample standard deviation.

(b) Find their coefficients of variation.

(c) Compare the mean, standard deviation, and CV of the two samples.

IV. Chebyshev's Theorem

Chebyshev's theorem tells us that for any set of data and for any constant k greater than 1, the proportion of the data that must lie within k standard deviations on either side of the mean is at least ___________________________________.

Focus Points
how to apply
Chebyshev's theorem
and what a Chebyshev
interval tells us

Example 6. For a sample with mean $\bar{x} = 5$ and standard deviation $s = 1.5$, find an interval A to B such that at least 75% of data will lie within this interval.

Example 7. For the sample in Example 6, find a Chebyshev interval about the mean in which at least 88.9% of the data fall.

Additional Notes

Homework Assignment

Page(s)

Exercises

Brase/Brase *Understanding Basic Statistics,* Sixth Edition, Student Notetaking Guide

Section 3.3 Percentiles and Box-and-Whisker Plots

<table>
<tr><td>Course Number</td></tr>
<tr><td>Instructor</td></tr>
<tr><td>Date</td></tr>
</table>

Objective:

In this section you learned how to interpret the meaning of percentile scores, how to compute the median, quartiles, and five-number summary from raw data, how to make a box-and-whisker plot and interpret results, and how a box-and-whisker plot indicates spread of data about the median.

Important Vocabulary   Define each term or concept.

Percentile

Quartiles

Interquartile range

Five-number summary

Box-and-whisker plot

I. Percentiles and Quartiles

Focus Points
how to interpret the meaning of percentile scores, how to compute the median and quartiles

There are ____ percentiles. If P is the Pth __________, then $P\%$ of data is less than or equal to P and $(100 - P)\%$ of data is greater than or equal to P. __________ are those percentiles that divide the data into fourths. The first quartile Q_1 is the ______ percentile, the second quartile Q_2 is the ________, and the third quartile Q_3 is the ______.

The difference between Q_3 and Q_1 is called the __________________.

Example 1. You took a test in your statistics class.

    (a) If your score is the 80th percentile, what percentage of scores are at or below yours?

    (b) If the score ranged from 1 to 100 and your raw score is 80, does this necessarily mean that your score is at the 80th percentile?

Example 2. For the sample: {10, 14, 11, 19, 15, 21, 21, 16, 20}

    (a) Find the quartiles.

    (b) Find the interquartile range.

Example 3. For the sample: {42, 77, 19, 53, 95, 34, 94, 86}

    (a) Order the data.

    (b) There are 8 data values. Find the median.

    (c) How many values are below the median position? Find Q_1.

    (d) There are the same number of data above as below the median. Use this fact to find Q_3.

    (e) Find the interquartile range and comment on its meaning.

II. Box-and-Whisker Plots

The smallest value, quartile, and the largest value together give us the ________________________ of the data and their spread.

A ____________________ is a visual representation a five-number summary, which contains a ______________ to include the lowest and highest data values. To the right of the scale, draw a _______ from Q_1 to Q_3. Include a solid line through the box at the _________ level. Draw vertical lines, called ____________, from Q_1 to the lowest value and from Q_3 to the highest value.

> **Focus Points**
> how to compute the five-number summary, how to make a box-and-whisker plot and how a box-and-whisker plot indicates spread of data about the median

Example 4. For the sample: {42, 77, 19, 53, 95, 34, 94, 86}

    (a) Find the five-number summary.

(b) Make a box-and-whisker plot.

(c) The box tells us that the middle half of the data lies
between ___ and ___, with an interquartile range ______.

(d) Is the median closer to the lower part of the box or the
upper part?

(e) Is the upper whisker longer or the lower whisker longer?
Does the data skew toward the smaller values or bigger values?

Example 5. Below are the box-and-whisker plots of the scores of a
test from two sections of the same course.

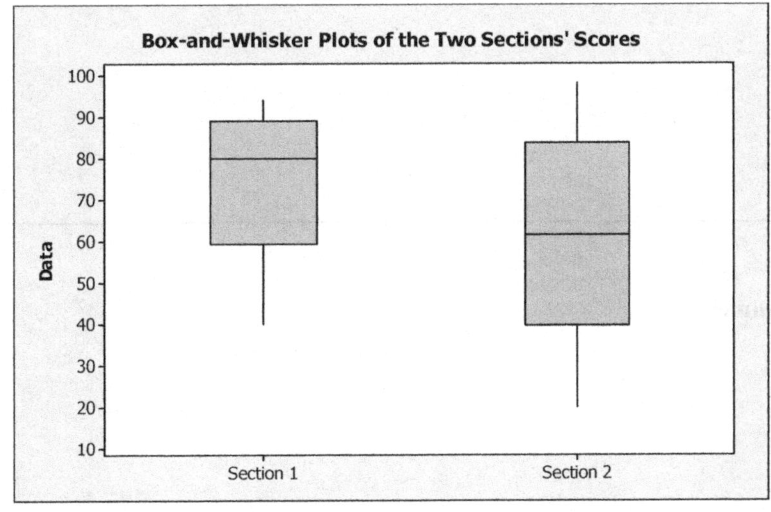

(a) Look at the plot of Section 1. Estimate the median and the extreme
values. In what range is the middle half of the scores?

(b) Look at the plot of Section 2. Estimate the median and the extreme values. In what range are the middle half of the scores?

(c) Compare the two plots and make comments about the scores of the two sections.

Additional Notes

Homework Assignment

Page(s)

Exercises

Chapter 4 Correlation and Regression

Section 4.1 Scatter Diagrams and Linear Correlation

| Course Number |
| Instructor |
| Date |

Objective:
In this section you learned how to make a scatter diagram, how to visually estimate the location of the "best-fitting" line, how to use sample data to compute the sample correlation coefficient r, and how to investigate the meaning of it.

Important Vocabulary   Define each term or concept.

Paired data values

Scatter diagram

Explanatory variable

Response variable

Sample correlation coefficient

Population correlation coefficient

I. Scatter Diagram

Studies of correlation and regression of _______ variables often involves a graph of paired data values _______.

Focus Points
how to make a scatter diagram, and how to visually estimate the location of the "best-fitting" line

In a _____________ , the data pairs (x, y) are plotted as individual points on a grid with _____________ axis x and _________ axis y. The first variable in the pair is called the ___________ , and the second one is called the _______________.

Example 1. For the paired data

x	12	17	18	20	25
y	18	19	19	20	21

(a) Make a scatter diagram along with a line segment showing the basic trend.

(b) Comment on the relation between x and y.

Example 2. For the paired data

x	1	2	3	4	5
y	8	9	7	5	4

(a) Make a scatter diagram.

(b) When the x value increases, what happens to the y values in general?

(c) Does a line fit the data reasonably well?

(d) Draw a line that you think "fits best."

II. Sample Correlation Coefficient *r*

The ______________________________ is a numerical value that measures the strength of a linear relationship between two variables.

> **Focus Point**
> the meaning of the correlation coefficient

Determine whether each of the following statements is true of false.

(a) The value of correlation coefficient *r* is always between –1 and 1.

(b) If *r* is positive, then *y* decreases as *x* increases.

(c) If we change the position of *x* and *y*, that is, plot (y, x) instead of (x,y), then the value of *r* will change.

(d) The value of *r* does not change when either *x* or *y* is converted into different units.

III. Development of the Formula for *r*

One formula for computing *r* is ______________________________.

> **Focus Point**
> how to compute the sample correlation coefficient *r*

IV. Computation Formula for *r*

Another formula for computing *r* is ______________________________.

Example 3. Continue from Example 1. For the paired data

x	12	17	18	20	25
y	18	19	19	20	21

(a) Do you expect to see a positive *r*?

(b) Compute the value of *r*.

Example 4. Continue from Example 2. For the paired data

x	1	2	3	4	5
y	8	9	7	5	4

(a) Do you expect to see a positive r?

(b) Compute the value of r.

Additional Notes

Homework Assignment

Page(s)

Exercises

Section 4.2 Linear Regression and the Coefficient of Determination

<table>
<tr><td>Course Number</td></tr>
<tr><td>Instructor</td></tr>
<tr><td>Date</td></tr>
</table>

Objective:

In this section you learned how to state the least-squares criterion, how to find the equation of the least-squares line, how to graph that line, how to use it to predict a value of the response variable for a specified value of the explanatory variable, how to explain the difference between interpolation and extrapolation, why extrapolation beyond the sample data range might give misleading or meaningless results, and how to use the coefficient of determination to determine explained and unexplained variation of the response variable.

Important Vocabulary Define each term or concept.

Least-squares criterion

Least-squares line

Interpolation

Extrapolation

Coefficient of Determination

Slope

Influential point

I. Least-squares Line

Focus Points
how to state the least-squares criterion and find the equation of the least-squares line, how to graph that line, and use it to make predictions, what is the difference between interpolation and extrapolation, why extrapolation might give misleading or meaningless results

Least-squares criterion states that the sum of the squares of the vertical distances from the data points (x, y) to the line must be made as _______ as possible. This line then is called the_____________________. The equation of this line is denoted by $\hat{y} = a + bx$ ($\hat{y}$ is also spelled as y-hat.) The value b is the _______ and a is the _________ of the line. Here $\hat{y}$ stands for the value of ___ estimated using the least-squares line and a given value of __.

Predicting values for x that are __________ observed x values in the data set is called interpolation. Predicting values of x values that are __________ observed x values in the data set is called extrapolation. __________________ may produce unrealistic forecasts.

Example 1. For the following paired data, find the equation of the least-squares line and graph it.

x	1	2	3	4	5
y	8	9	7	6	4

(a) Find the slope b and the y-intercept a.

(b) Graph the least-squares line on a scatter diagram.

Example 2. Continue with Example 1.

(a) In Example 1, which variable is the explanatory variable? Which is the response variable?

(b) Write the equation of the least-squares line. Can you use the equation to predict the y value for a specified x value?

(c) The sample data pairs have x values ranging from 1 to 5. To predict y when x equals 3.5, do you interpolate or extrapolate?

(d) Predict y when x equals 3.5.

Example 3. Given the data

x	10	20	30	40	50
y	21	23	25	23	27

(a) Draw a scatter diagram for the data.

(b) Find $\sum x, \sum y, \sum x^2$, and $\sum xy$.

(c) Find $\bar{x}$ and $\bar{y}$.

(d) Compute the value of a and b.

(e) What is the equation of the least-squares line?

(f) Plot the least-squares line on the scatter diagram.

(g) Approximate the y value for $x = 25$ from the graph. Then use the equation of the least-squares line to calculate y when $x = 25$.

II. Coefficient of Determination

The square of the sample correlation coefficient r is called the
__________________. It stands for the ratio of explained variation
over total variation.

Focus Points
how to use the coefficient of determination to determine explained and unexplained variation of the response variable

Example 4. Continue with Example 3

   (a) Find the correlation coefficient r for that data.

   (b) Compute the coefficient of determination.

   (c) What percentage of the variation in y can be explained
by x and the least-squares line?

   (d) What percentage of the variation in y cannot be explained
by x and the least-squares line?

Additional Notes

Homework Assignment

Page(s)

Exercises

Chapter 5 Elementary Probability Theory

Section 5.1 What is Probability?

Course Number

Instructor

Date

Objective:
In this section you learned how to assign probabilities to events, how the law of large numbers relates to relative frequency, how to apply basic rules of probability in every day life, and how to explain the relationship between statistics and probability.

Important Vocabulary   Define each term or concept.

Probability

Law of large numbers

Statistical experiment

Event

Simple event

Sample space

Complement of an event

I. Probability

Focus Points
how to assign probabilities to events, how the law of large numbers relates to relative frequency

______________ is used to state the chances for an event to occur. The value of a probability is always between ___ and ___. The notation ______ denotes the probability of event A. When $P(A)$ is closer to 1, then the event A is ________ likely to occur. If $P(A)$ is closer to 0, then the event A is _______ likely to occur.

The law of large numbers states that in the long run, as the sample size gets bigger and bigger, the ____________________ of outcomes will become closer and closer to the actual ____________________ value.

Brase/Brase *Understanding Basic Statistics*, Sixth Edition, Student Notetaking Guide   **41**

A statistical experiment or statistical observation can be considered to be any _________ activity that results in a _____________ outcome.

An _____________ is a group of outcomes in a statistical experiment. A _____________ event is one particular outcome of a statistical experiment. The complement of event A, denoted by A^c, is the event that _________ _________. The set of all simple events is called the _______________ of an experiment.

Example 1. Consider each of the following events, and determine how the probability is assigned.

   (a) Lisa thinks she has a 95% chance of passing the history course.

   (b) If you toss a fair coin, then the probability of getting a Head is 0.5.

   (c) An insurance company claims that the probability of getting an erroneous medical expense report is 0.2, based on a random sample of 400 reports, of which 80 were erroneous.

Example 2. Assign a probability to each following event based on the information provided.

   (a) A random sample of 400 students at a college were surveyed and 300 of them liked the food in the cafeteria. Estimate the probability that a student randomly selected at that college likes the cafeteria food.
   (b) Toss a fair die. What is the probability that you get a number less than or equal to 4?
   (c) John is to take a driver's license test. Based on his knowledge and skills, John is almost sure that he will pass the test. What would he say about his probability of passing?

Example 3. Randomly toss two fair coins. What is the sample space of this experiment? What is the probability that you get two Heads? What is the probability that you get at least one Head?

Example 4. Randomly throw a die. What is the sample space of this experiment? What is the probability that you get an even number? A number greater than 4?

Example 5. If $P(A) = 0.35$, what is the value of $P(A^c)$?

II. Probability Related to Statistics

Probability states what will occur when samples are drawn from a _____ population, while statistics describes how samples are to be obtained and how inferences are to be made about __________ populations.

Focus Points
how to apply probability rules in every day life, and how to explain the relationship between statistics and probability

Example 6. Determine whether each of the following examples is a probability application or a statistical application.

   (a) Randomly select two students from a class which has 14 boys and 18 girls. What is the probability that both selected students are boys?
   (b) Randomly interview 100 high school students and ask them how much time they spent on sports each week and find their average. Based on the sample results, make a conjecture about the average amount of time a high school student would spend on sports per week.

Additional Notes

Homework Assignment

Page(s)

Exercises

Section 5.2 Some Probability Rules – Compound Events

Objective:
In this section you learned how to compute probabilities of general compound events, how to compute probabilities involving independent events or mutually exclusive events, and how to use survey results to compute conditional probabilities.

Course Number

Instructor

Date

Important Vocabulary   Define each term or concept.

Independent events

Dependent events

Conditional probability

Multiplication rules

Mutually exclusive events

Addition rules

Basic probability rules

I. Conditional Probability and Multiplication Rules

Focus Points
how to compute probabilities of general compound events, how to compute probabilities involving independent events

Two events A and B are ________________ if the occurrence or nonoccurrence of one does not change the probability that the other will occur. If events are ______________, the probability of one event depends upon the occurrence of the other event.

________________________ is the probability that a dependent event will occur given that another event has occurred.

Multiplication rules state that if events A and B are ____________, then $P(A \text{ and } B) = P(A)P(B)$; if A and B are ______________ then $P(A \text{ and } B) = P(A)P(B|A) = P(B)P(A|B)$.

Example 1. Suppose you are going to throw two fair dice. What is the probability of getting a 3 on each die?

Example 2. Suppose you are drawing two cards without replacement from a well-shuffled deck of 52 cards. What is the probability of getting two heart cards?

Example 3. John and Lisa are going to take the driver's license test. Based on their skills, the probability that John will pass the test is 0.7 and the probability that Lisa will pass is 0.8. Assuming that the test result of one of them will have no effect on the result of the other, answer these two questions.

   (a) Are the events *John will pass* and *Lisa will pass* dependent or independent?
   (b) What is the probability that they both will pass?

Example 4. Suppose you are drawing two cards without replacement from a well-shuffled deck of 52 cards.

   (a) What is the probability of getting a king on the first draw?
   (b) Suppose the first draw is indeed a king. What is the probability of getting a king again on the second draw?
   (c) Are the answers to (a) and (b) different?
   (d) What is the probability that both draws yield kings?

II. Additional Rules

If two events cannot occur together, then they are ___________________.
In particular, events A and B are mutually exclusive if $P(A \text{ and } B) =$ __.
For mutually exclusive events A and B, the probability $P(A \text{ or } B) =$
___________. If the events are not mutually exclusive, then $P(A \text{ or } B)$
= ___________________.

Focus Points
how to compute probabilities of general compound events, how to compute probabilities involving mutually exclusive events

Example 5. Indicate how each of the following pairs of events are combined. Use either the *and* combination or the *or* combination.

   (a) Taking a vacation either in New York City or in Washington D.C.
   (b) Taking a vacation to see both New York City and Washington D.C.
   (c) Completing the major requirements not only in mathematics but also in computer science.
   (d) Completing the major requirements in at least one of these majors: mathematics, computer science, statistics.

Example 6. Suppose you are drawing one card from a well-shuffled deck of 52 cards. Consider these two events: A: you get a heart card; B: you get a spade card.

   (a) Are the events A and B mutually exclusive?
   (b) What is the probability that you get either a heart card or a spade card?

Example 7. Suppose you are drawing one card from a well-shuffled deck of 52 cards. Consider these two events: A: you get a king; B: you get a spade card.

   (a) Are the events A and B mutually exclusive?
   (b) What is the probability that you get either a king or a spade card?

Example 8. Suppose you are throwing two fair dice. What is the probability that you will roll a sum bigger than 10?

III. Further Examples

A ________________ is a survey that is formed by questions to which the responses can be recorded in a table called a ________________ .

> *Focus Points*
> how to use survey results to compute probabilities

Example 9. All 7000 students at a college were surveyed about their residency status(live on campus, or live off campus but in town, or commute). The students were grouped by gender. The results are shown in the following contingency table.

	On Campus	Off Campus	Commute	Row Total
Male Students	1700	1300	600	3600
Female Students	1800	1200	400	3400
Column Total	3500	2500	1000	7000

Suppose a student is selected at random from these 7000 students. Consider these events: M = male student; F = female student; C = on campus; O = off campus; T = commute

   (a) Find $P(C)$ and $P(M)$.

     Brase/Brase *Understanding Basic Statistics,* Sixth Edition, Student Notetaking Guide

(b) Find $P(C|M)$.

(c) Are the events C and M independent?

(d) Find $P(C \text{ and } M)$.

(e) Find $P(C \text{ or } M)$.

Example 10. Continue with Example 9.

(a) Find $P(O)$ and $P(F)$.

(b) Find $P(O|F)$.

(c) Find $P(O \text{ and } F)$.

(d) Use the multiplication rule to find $P(O \text{ and } F)$. Is the result

the same as in (c)?

(e) Find $P(O \text{ or } F)$. Are the events mutually exclusive?

Additional Notes

Homework Assignment

Page(s)

Exercises

Section 5.3 Trees and Counting Techniques

<table>
<tr><td>Course Number</td></tr>
<tr><td>Instructor</td></tr>
<tr><td>Date</td></tr>
</table>

Objective:
In this section you learned how to organize outcomes in a sample space using tree diagrams, how to compute the number of ordered arrangements of outcome, how to compute the number of nonordered groupings of outcomes, and how counting techniques relate to probability in every-day life.

Important Vocabulary Define each term or concept.

Multiplication rule of counting

Tree diagram

Factorial notation

Permutation

Combination

I. Multiplication Rule of Counting and Tree Diagram

Focus Points
how to organize outcomes in a sample space using tree diagrams and how to use the multiplication rule of counting

The multiplication rule of counting states that the total number of possible outcomes for a sequence of events equals the ______ of the number of possibilities for each event in the sequence.

To display of the total number of outcomes of an experiment consisting of a series of events, you may use a ______ diagram.

Example 1. There three reference books, five cookbooks, and four novels on the shelf. Jeff is to pick one book of each kind. How many different sets are possible?

  Brase/Brase *Understanding Basic Statistics,* Sixth Edition, Student Notetaking Guide

Example 2. Suppose you are tossing two fair coins. Use a tree diagram
to show all possible outcomes.

Example 3. Continue with Example 2.

    (a) How many branches are there at the end of the tree diagram?
How many outcomes are there in this experiment?

    (b) Complete the list of all outcomes.

1st coin	2nd coin
H	H
H	T
___	___
___	___

    (c) Use the multiplication rule to compute the total number of outcomes
in this experiment.

Example 4. Suppose you are drawing two cards without replacement
from a well-shuffled deck of 52 cards. You are to note the color of the first
card (red or black), and then note the color of the second card.

    (a) What are the outcomes of this experiment?

    (b) What is the probability of each outcome?

II. Factorial and Permutation

> *Focus Points*
> how to compute the
> number of ordered
> arrangements
> of outcomes

For a number $n>0$, its factorial $n! = $ ________________________. By special
definition, $0! = $ ____.

The number of different ways of taking r objects out of n objects
considering the order, called the number of ____________ of r out of n,
is denoted by _______.

The formula for computing the number of permutations is: $P_{n,r} = $

________________.

Example 5.

    (a) Evaluate 5!

    (b) In how many different ways can five objects be arranged in order? How many choices do you have for the first position? For the second position? For the third, fourth and fifth position?

Example 6. Compute the number of possible ordered seating arrangements for nine people in six chairs.

III. Combination

The number of different ways of taking r objects out of n objects not considering the order, called the number of ______________ of r out of n, is denoted by ________.

The formula for computing the number of combinations is: $C_{n,r} =$ ______________.

> *Focus Points*
> how to compute the number of nonordered groupings of outcomes

Example 7. There are eight different books on the shelf. How many different groups of three books can be selected from the shelf?

Example 8. From a student organization of 20 members, three officers – president, vice president, and treasurer – must be elected. How many different slates of officers are possible?

    (a) Are the slates permutations or combinations?

    (b) How many different slates are possible?

Example 9. From a student organization of 20 members, three members will be selected to attend a convention. How many different groups of 3 are there?

    (a) Are the slates permutations or combinations?

    (b) How many different groups are possible?

Additional Notes

<table>
<tr><td>

Homework Assignment

Page(s)

Exercises

</td></tr>
</table>

Chapter 6 The Binomial Probability Distribution and Related Topics

Section 6.1 Introduction to Random Variables and Probability Distribution

Course Number

Instructor

Date

Objective:

In this section you learned how to distinguish between discrete and continuous random variables, how to graph discrete probability distributions, and how to compute μ and σ for a discrete probability distribution.

Important Vocabulary   Define each term or concept.

Random variable

Discrete random variable

Continuous random variable

Probability distribution

Mean of a discrete probability distribution

Standard deviation of a discrete probability distribution

Expected value

I. Random Variable

Focus Points
how to distinguish
between discrete and
continuous random
variables

A quantitative variable x is a _______________ if the value that x takes on is a random outcome from a given experiment. If x can take on only a finite number of values or a countable number of values, then x is a __________ random variable. If x can take on any value in a line interval, then it is a _____________ random variable.

Example 1. Which of the following random variables are discrete and which are continuous?

    (a) the time it takes for a boy to run 100 meters

    (b) the number of pencils left on a table

    (c) the amount of gasoline left in a tank

    (d) the number of history majors in a random sample of 100 college students

II. Probability Distribution of a Discrete Random Variable

An assignment of probabilities to each distinct value of a __________ random variable or to each interval of values of a ______________ random variable is called a __________________.

Focus Points
how to graph discrete probability distributions, how to compute μ and σ for a discrete probability distribution

The __________ of a discrete probability distribution if the a "central point" or "cluster point" for the entire distribution. It is also called the __________________. The ____________________ is a measure of the likelihood that the variable x is different from the mean.

Example 2. A quiz with possible scores 1, 2, 3, and 4 was given to a class of 20 students. The score distribution is listed in this table:

Score x	Number of students
1	3
2	7
3	6
4	4

    (a) If you randomly select a student from this class, what is the probability that his or her score is 2?

    (b) Make a probability distribution table for this distribution.

(c) Make a graph for this probability distribution.

(d) Find $P(3 \text{ or } 4)$.

Example 3. The same quiz as in Example 2, with possible scores 1, 2, 3, and 4, was also given to another class of 25 students.

(a) Part of the score distribution and corresponding probability distribution table is given below. Complete the table.

Score x	Number of students	Probability
1		0.2
2	6	___
3	10	___
4	___	0.16

(b) Do the probabilities of all the scores add up to 1?

(c) If a student is selected from this class at random, what is the probability that his or her score is below 3?

Example 4. For the probability distribution in Example 2, find the mean and standard deviation.

Example 5. At a carnival, you pay $1.00 to play a coin-flipping game with two fair coins. If you get Head on both coins, then you win $3.00. Are your expected earnings equal to the cost to play?

(a) What is the random variable of interest in this case? What is the sample space for this random variable?

(b) There are four equally likely outcomes for throwing two coins. What are they?

(c) Complete the following table regarding the earning.

Earning x	Frequency	$P(x)$	$xP(x)$
0	3	___	___
3	___	0.25	___

(d) Find the expected earnings. Is that more than, equal to, or less than the cost to play? What does that mean in the long run?

Additional Notes

Homework Assignment

Page(s)

Exercises

Section 6.2 Binomial Probabilities

<table>
<tr><td>Course Number</td></tr>
<tr><td>Instructor</td></tr>
<tr><td>Date</td></tr>
</table>

Objective:

In this section you learned how to recognize a binomial experiment, how to compute binomial probabilities using the formula $P(r)=C_{n,r}p^{r}q^{n-r}$, how to use the binomial table to find $p(r)$, and how to use the binomial probability distribution to solve real-world applications.

Important Vocabulary   Define each term or concept.

Binomial experiment

Binomial coefficient

Binomial probability distribution

Number of trials

Probability of successes and failures

I. Binomial Experiment

Focus Points
how to recognize a binomial experiment

The problem to be solved in a ____________ experiment is to find the probability of r successes out of n trials. Each trial is independent and has one of two outcomes, ________ or _________.

Example 1. Consider this problem: throw a fair die 10 times, and find the probability that the number 3 appears twice. Let's see if this problem fits the definition of a Binomial experiment.

(a) Is there a fixed number, n, of trials? If so, what is n?

(b) Are the trials independent?

(c) How many outcomes are there in each trial? What are they?

(d) In each trial, what is the probability for S for F?

(e) We like to know the probability that 3 appears for 2 times in 10 trials. Is this a Binomial experiment? If so, what are the values of r, p, q?

Example 2. In a certain college, 60% of students live on campus. Suppose we randomly choose 20 students from this college and ask whether each of them live on campus. What is the probability that 5 of them live on campus?

(a) If a student lives on campus, we say that we have a success. With this definition for success, what is failure?

(b) Since 60% of students live on campus, the probability for success is 0.6. What is the probability for failure, q?

(c) In this experiment, there are $n = $ _______ trials.

(d) We wish to find out the probability that 5 of these 20 students live on campus. In this case, $r = $ ______.

II. Computing Probabilities for a Binomial Experiment Using the Binomial Distribution Formula

Focus Points
how to compute binomial probabilities using the formula
$P(r) = C_{n,r} p^r q^{n-r}$

The ___________________ equals $C_{n,r}$. The _______________ probability distribution can be used to compute the probability of r successes for n of trials. The formula is $P(r) = $ ______________.

Example 3. Use the binomial distribution formula to find the probability described in Example 1.

(a) In this case, what are the values of n, p, q, and r?

(b) Apply the formula $P(2) = C_{10,2}(1/6)^2(5/6)^{10-2}$ to find the value of $P(2)$.

III. Computing Probabilities for a Binomial Experiment Using a Binomial Distribution Table

Focus Points
how to compute binomial probabilities using a binomial distribution table

Example 4.

(a) Use the binomial distribution table given in the text to find the probability described in Example 2.

(b) Continue with Example 2. What is the probability that no more than 10 of them live on campus?

Example 5. A college conducts a survey each year to see whether students are happy with the cafeteria service. Last year only 40% of the students were happy with it. The cafeteria has made a number of changes and improvements this year. A random interview with 10 students shows that 7 out of 10 are happy this year. Do the interview results indicate that the cafeteria is doing better this year?

(a) Each individual interview is a binomial trial. In this case, $n =$ _____,
$p =$ _____, $q =$ _____, $r =$ _____.

(b) Use either the formula or the table to find $P(7)$.

(c) Use either the formula or the table to find $P(r \geq 7)$.

(d) Does the cafeteria seem to be doing better this year?

Additional Notes

Homework Assignment

Page(s)

Exercises

 Brase/Brase *Understanding Basic Statistics,* Sixth Edition, Student Notetaking Guide

Section 6.3 Additional Properties of the Binomial Distribution

Course Number

Instructor

Date

Objective:
In this section you learned how to make histograms for binomial distributions, and how to compute μ and σ for a binomial distribution.

Important Vocabulary   Define each term or concept.

Mean of a binomial distribution

Standard deviation of a binomial distribution

I. Graphing a Binomial Distribution

Focus Points
how to make histograms
for binomial distributions

Example 1. Make a histogram for a binomial distribution with $n = 5$ and $p = 0.4$.

Example 2. Suppose you are to throw a fair coin 4 times. Make a histogram showing the probability that the Head will appear 0, 1, 2, 3, and 4 times.

(a) This is a binomial experiment with $n =$ _____ and $p =$ _____.

(b) Use the table given in the text to make a distribution table for this binomial distribution.

(c) Make a histogram for this distribution.

(d) What is the area of the bar over $r=1$? What is the area of the bar over $r = 3$? How does the probability that you get 3 heads compare with the probability that you get 1 head?

II. Mean and Standard Deviation of a Binomial Distribution

The mean for a binomial distribution can be computed by the formula $\mu =$ _______, and the standard deviation can be computed by the formula $\sigma =$ _________ .

Focus Points
how to compute μ and σ for a binomial distribution

Example 3. For a binomial distribution with $n = 5$ and $p = 0.4$, find μ and σ.

Example 4. Continue with Example 2. Find μ and σ for the distribution described.

Homework Assignment

Page(s)

Exercises

     Brase/Brase *Understanding Basic Statistics,* Sixth Edition, Student Notetaking Guide

Chapter 7 Normal Curves and Sampling Distributions

Section 7.1 Graphs of Normal Probability Distributions

Objective:
In this section you learned how to graph a normal curve and summarize its important properties, and how to apply the empirical rule to solve real-world problems.

<table>
<tr><td>Course Number</td></tr>
<tr><td>Instructor</td></tr>
<tr><td>Date</td></tr>
</table>

Important Vocabulary   Define each term or concept.

Normal distribution

Normal curve

Symmetry of a normal curve

Empirical rule

I. Normal Distribution and Normal Curve

A __________ distribution is one of the most important examples of a continuous probability distribution . The graph of a normal distribution is called a ________ curve.

Focus Points
how to graph a normal
curve and summarize
its important properties

Example 1. Look at the normal curves below.

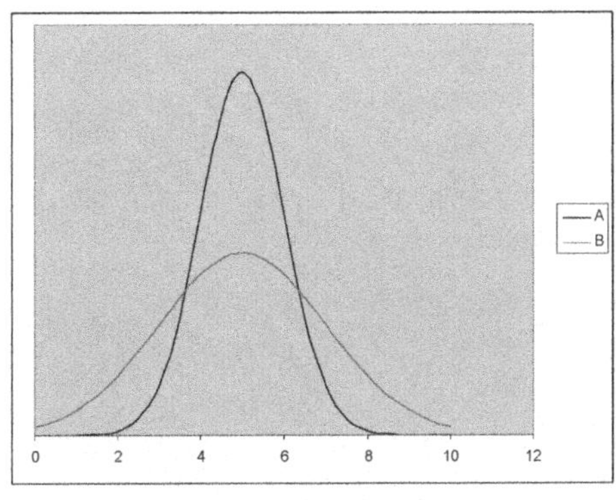

(a) Do these two distributions have the same mean? If so, what is it?

(b) One of the curves has a standard deviation $\sigma = 1$ and the other has $\sigma = 2$. Which curve has which σ?

II. Empirical Rule

The ______________ rule states that for a distribution that is symmetrical and bell-shaped (in particular, for a normal distribution), about ______ of the data values are within one standard deviation from the mean, about ______ of the data values are within two standard deviations from the mean, and about ______ of the data values are within three standard deviations from the mean.

> **Focus Points**
> how to apply the empirical rule to solve real-world problems

Example 2. The scores of a recent SAT test are normally distributed with mean $\mu = 1100$ and standard deviation $\sigma = 200$. What is the probability that a score selected at random will be between 1100 and 1500?

Example 3. Continue with Example 2.

(a) Is the probability that the score is between 700 and 1100 the same as the probability that the score is between 1100 and 1500?

(b) What is the probability that the score is between 1100 and 1300?

(c) What is the probability that the score is between 1300 and 1500?

Homework Assignment

Page(s)

Exercises

Brase/Brase *Understanding Basic Statistics,* Sixth Edition, Student Notetaking Guide

Section 7.2 Standard Units and Areas Under the Standard Normal Distribution

Objective:
In this section you learned how to convert raw data to z scores with given μ and σ, how to convert z scores to raw data with given μ and σ, how to graph the standard normal distribution, and how to find areas under the standard normal curve.

Course Number

Instructor

Date

Important Vocabulary Define each term or concept.

z score

Raw score

Standard normal distribution

Area under a standard normal curve

I. z Scores and Raw Scores

The __ score or the __ value tells how many standard deviations is the _____ score x from the mean of the x distribution.

Focus Points
how to convert raw data to z scores with given μ and σ, how to convert z scores to raw data with given μ and σ

Example 1. Suppose $\mu = 2$ and $\sigma = 3$. What is the z score of the raw score = 6.5?

Example 2. Suppose $\mu = -1$ and $\sigma = 2$. What is the z score of the raw score = 0? What raw score corresponds to a z score -2?

II. Standard Normal Distribution

The standard normal distribution is a normal distribution with mean __ and standard deviation __.

Focus Points
how to describe a standard normal distribution

III. Areas Under the Standard Normal Curve

To find areas under the standard normal curve, we use a _____________ table. This table gives cumulative areas to the _____ of a specified z.

Focus Points
how to graph the standard normal distribution, and how to find areas under the standard normal curve

Example 3 Use the table given in the text to find

(a) the area under the standard normal curve to the left of $z = -1.23$.

(b) the area under the standard normal curve to the left of $z = 2.1$.

Example 4.

(a) If $z < 0$, is the area to the left of z less than 0.5 or greater than 0.5?

(b) If $z > 0$, is the area to the left of z less than 0.5 or greater than 0.5?

(c) As z values decrease, do the areas to the left of z increase?

Example 5.

(a) Find the area between $z = 0.5$ and $z = 1.5$

(b) Find the area to the right of $z = 0.3$.

Example 6. Let z be a random variable with a standard normal distribution.

(a) Find the probability $P(z > 1.89)$.

(b) Find the probability $P(-1 < z < 0.5)$.

Homework Assignment

Page(s)

Exercises

Section 7.3 Areas Under Any Normal Curve

Course Number

Instructor

Date

Objective:

In this section you learned how to compute the probability of "standard events," how to find a z score from a given normal probability (inverse normal), and how to use the inverse normal to solve guarantee problems.

Important Vocabulary   Define each term or concept.

Inverse normal probability distribution

Pearson's index

Normal quantile plot (or normal probability plot)

I. Normal Distribution Areas

Focus Points
how to compute probability from a normal distribution

Example 1. Suppose x has a normal distribution with $\mu = 2$ and $\sigma = 3$. Find $P(-1 < x < 8)$.

Example 2. Suppose x has a normal distribution with $\mu = -1$ and $\sigma = 2$.

(a) Convert the statement $x \leq 2$ to a statement about z.

(b) What is the area under the standard normal curve to the left of $z = 1.5$?

(c) What is the probability $P(z \leq 1.5)$?

(d) What is the probability $P(x \leq 2)$?

II. Inverse Normal Distribution

Focus Points
how to find a x or z value from a given normal probability, and how to check a data for normality

Suppose x has a normal probability distribution. When we try to find the x value that corresponds to a given area under the normal curve, we use the __________ normal probability distribution.

Pearson's index can be used to test if data have a normal distribution. Normal distributions are symmetric and should have a Pearson's index value between ___ and ___.

A ________________ can also be used to test if data have a normal distribution. If the data have a normal distribution, then the normal quantile plot of the data should be close to a __________ line.

Example 3. Suppose x has a normal distribution with $\mu = -1$ and $\sigma = 2$. Find the value of c such that $P(x < c) = 0.512$.

Example 4. Find the z value such that 80% of the area under the standard normal curve lies between $-z$ and z.

Example 5.

(a) Find the z value such that 60% of the area under the standard normal curve lies to the right of z.

(b) Suppose x has a normal distribution with $\mu = 1$ and $\sigma = 2$. Find the x value such that 60% of the area under the standard normal curve lies to the right of x.

Example 6. Consider the data $\{1, 5, 2, 3, 4, 3, 3, 4, 4, 3, 2\}$

(a) Make a histogram and box-and-whisker plot for this data.

 Brase/Brase *Understanding Basic Statistics,* Sixth Edition, Student Notetaking Guide

(b) Use Pearson's index to check the normality.

(c) Make a normal quantile plot for this data.

(d) Interpret the results.

Additional Notes

Homework Assignment

Page(s)

Exercises

 Brase/Brase *Understanding Basic Statistics,* Sixth Edition, Student Notetaking Guide

Section 7.4 Sampling Distributions

Course Number

Instructor

Date

Objective:

In this section you reviewed terms such as random sample, relative frequency, parameter, statistic, and sampling distribution. You also learned how to construct a relative frequency distribution for $\bar{x}$ values from raw data, and how to compare the result to a theoretical sampling distribution.

Important Vocabulary   Define each term or concept.

Estimation

Testing

Regression

Sampling distribution

I. Sampling Distributions

Focus Points
how to construct a relative frequency distribution for $\bar{x}$ values from raw data, and how to compare the result to a theoretical sampling distribution

When we estimate the value of a population parameter, we are making an ___________ . When we formulate a decision about the value of a population parameter, we are ___________. When we make predictions or forecasts about the value of a statistical variable, we are doing a _________________.

A _________________ is a probability distribution of a sample statistic based on all possible simple random samples of the ______ size from the same population.

Example 1. Suppose that 50 randomly selected groups, each containing 10 high school seniors, were interviewed. For each group, the average SAT score $\bar{x}$ of the 10 students was calculated and recorded. It was found that among these 50 average scores, 3 were between 400 and 600, 5 were between 600 and 800, 16 were between 800 and 1000, 17 were between 1000 and 1200, 5 were between 1200 and 1400, and 4 were between 1400 and 1600.

(a) Make a relative frequency distribution table for the average score $\bar{x}$.

(b) Make a relative frequency histogram for the average score $\bar{x}$.

Homework Assignment

Page(s)

Exercises

Brase/Brase *Understanding Basic Statistics,* Sixth Edition, Student Notetaking Guide

Section 7.5 The Central Limit Theorem

<table>
<tr><td>Course Number</td></tr>
<tr><td>Instructor</td></tr>
<tr><td>Date</td></tr>
</table>

Objective:

In this section you learned how to use μ and σ to construct the theoretical sampling distribution for $\bar{x}$ for a normal distribution, how to use sample estimates to construct a good approximate sampling distribution for $\bar{x}$ for large samples, and how to explain the meaning of the Central Limit Theorem.

Important Vocabulary   Define each term or concept.

Standard error

Central Limit Theorem

Unbiased

Variability

Large Sample

I. The $\bar{x}$ Distribution, Given x Is Normal

Focus Points
how to use μ and σ to construct the theoretical sampling distribution for $\bar{x}$ for a normal distribution

The standard error is another name for the _______________ of the $\bar{x}$ distribution.

Let x be a random variable that has a normal distribution with mean μ and standard deviation σ. Let $\bar{x}$ be the sample mean corresponding to random samples of size n taken from the x distribution. Then the following are true:

(a) $\bar{x}$ has a _________ distribution.

(b) The mean of the $\bar{x}$ distribution equals ___.

(c) The standard deviation of the $\bar{x}$ distribution equals _____

Example 1. Suppose x has a normal distribution with $\mu = 1$ and $\sigma = 2$.

(a) Find the probability $P(0 < x < 4)$.

(b) Let $\bar{x}$ be the sample mean corresponding to random samples of size 16 taken from the x distribution. Find the probability $P(\,0 < \bar{x} < 4)$.

(c) Are the answers of (a) and (b) quite different? Why is this the case?

II. The $\bar{x}$ Distribution, Given x Follows Any Distribution

The central limit theorem says that no matter what distribution x follows, when the sample size gets larger and larger, the distribution of $\bar{x}$ will approach a __________ distribution. It is commonly agreed that if the sample size is equal to or greater than ____, then the $\bar{x}$ distribution will appear normal and the central limit theorem will apply.

> **Focus Points**
> how to use sample estimates to construct a good approximate sampling distribution for $\bar{x}$ for large samples, and how to explain the meaning of the Central Limit Theorem

A sample statistic is unbiased if the _______ of its sampling distribution _______ the value of the parameter being estimated.

The spread of the sampling distribution indicates the __________ of the statistic.

Example 2. A certain kind of bacteria exists in all water. Let x be the bacteria count per milliliter of water. The health department has found that if the water is not contaminated, then x has a distribution that is more or less mound-shaped and symmetrical. The mean of x is $\mu = 3000$ and the standard deviation is $\sigma = 200$. The city health inspector takes 64 random samples of water from the city public water system each day. The bacteria count in each of the 64 samples is averaged to obtain the sample mean bacteria count $\bar{x}$.

(a) Assuming the public water is not contaminated, what is the distribution of $\bar{x}$? What is the mean of $\bar{x}$? What is the standard deviation of $\bar{x}$?

(b) Assuming the public water is not contaminated, what is the probability that $\bar{x}$ is between 2925 and 3075?

(c) Suppose one day the mean bacteria count $\bar{x}$ is not between 2925 and 3075. If you were the inspector, what would be your comment on this situation?

Example 3. Continue with Example 2.

    (a) What theorem supports your answer to the first question of part (a) of Example 2?

    (b) Find $P(2950 < \bar{x} < 3050)$.

    (c) Interpret your answer to part (b).

Additional Notes

<table>
<tr><td>

Homework Assignment

Page(s)

Exercises

</td></tr>
</table>

Section 7.6 Normal Approximation to the Binomial Distribution and to the $\hat{p}$ Distribution

Course Number

Instructor

Date

Objective:
In this section you learned the assumptions needed to use the normal approximation to the binomial distribution. You also learned how to compute μ and σ for the normal approximation, how to use the continuity correction to convert a range of r values to a corresponding range of normal x values, and how to convert the x values to a range of standardized z scores and find desired probabilities. You also learned to describe the sampling distribution for proportions $\hat{p}$.

Important Vocabulary Define each term or concept.

Normal approximation to the binomial distribution

Continuity correction

Sampling distribution for $\hat{p}$

Standard error of a proportion

---

1. Normal Approximation to the Binomial Distribution

Suppose r has a binomial distribution with number of trials n and probability of success on a single trial p. Also let $q = 1 - p$. If $np > \underline{\hspace{1cm}}$ and $nq > \underline{\hspace{1cm}}$, then r has approximately a $\underline{\hspace{2cm}}$ distribution with $\mu = \underline{\hspace{1cm}}$ and $\sigma = \underline{\hspace{1.5cm}}$.

> **Focus Points**
> when to use the normal approximation to the binomial distribution, how to compute μ and σ for the normal approximation, and how to use the continuity correction to convert a range of r values to a corresponding range of normal x values

Continuity correction needs to be made when approximating a binomial distribution with a normal distribution. If the discrete variable is a left point of an interval, $\underline{\hspace{2cm}}$ 0.5 to obtain the corresponding normal variable. If the discrete variable is a right point of an interval, $\underline{\hspace{1.5cm}}$ 0.5 to obtain the corresponding normal variable.

    Brase/Brase *Understanding Basic Statistics*, Sixth Edition, Student Notetaking Guide

Example 1. Graph the binomial distributions with $p = 0.3$, and

(a) $n = 5$.

(b) $n = 30$

Example 2. Continue with Example 1.

(a) Look at the graph you obtained for part (a) of Example 1.
Is it reasonable to approximate that distribution with a
normal distribution? What are the values of np and nq in
this case?

(b) Look at the graph you obtained for part (b) of Example 1.
Is it reasonable to approximate that distribution with a normal
distribution? What are the values of np and nq in this case?

(c) Continue with part (b) of Example 1. Suppose we are using
the normal distribution x to approximate the probability
$P(7 \leq r \leq 10)$. What should be the mean and standard deviation

of that normal distribution? What is the range of x after the continuity correction?

(d) Use the results of part (c) to approximate $P(7 \leq r \leq 10)$ with that normal distribution. What is that approximate value of $P(7 \leq r \leq 10)$?

(e) Use the binomial distribution to find the exact value of $P(7 \leq r \leq 10)$.

(f) How do the results of parts (d) and (e) compare? Is the difference negligible?

Example 3. Throw a fair coin for 200 times and let r be the number of Heads occurring in the 200 trials. Since the coin is fair, $p = 0.5$ and $q = 0.5$.

(a) To approximate $P(90 \leq r \leq 120)$, we use the normal curve with $\mu =$ _____ and $\sigma =$ ______ .

(b) $P(90 \leq r \leq 120)$ is approximately equal to $P(\_\_\_\_\_\_ \leq x \leq \_\_\_\_\_)$, where x is a variable from the normal distribution described in part (a).

(c) Convert the probability about x in part (b) into a probability about the standard normal distribution z: $P(\_\_\_\_\_\_ \leq z \leq \_\_\_\_\_)$.

(d) Find that approximate value using the result of part (c).

(e) Will the normal distribution make a good approximation to the binomial for this problem? Explain your answer.

Additional Notes

Homework Assignment

Page(s)

Exercises

Chapter 8 Estimation

Section 8.1 Estimating μ When σ Is Known

Course Number

Instructor

Date

Objective:
In this section you learned how to explain the meaning of confidence level, error of estimate, and critical value, how to find the critical value corresponding to a given confidence level, how to compute confidence intervals for μ when σ is known, how to interpret the results, and how to compute the sample size to be used for estimating a mean μ.

Important Vocabulary   Define each term or concept.

Point estimate

Margin of error

Confidence level

Critical value

Confidence interval for μ

I. Confidence Interval For μ When σ Is Known

Focus Points
how to explain the meaning of confidence level, error of estimate, and critical value, how to find the critical value corresponding to a given confidence level, how to compute confidence intervals for μ when σ is known and interpret the results

A ________________ of a population parameter is an estimate of the parameter using a single number. A sample ______ is a point estimate of the population mean.

The ________________ is the absolute value of the difference between the sample point estimate and the true population parameter value.

The ______________ stands for the reliability of an estimate. Usually c is chosen to be a number such as ____, ______, or ______.

For a confidence level c, the __________ z_c is the value such that the area under the standard normal curve between $-z_c$ and z_c equals c.

   Brase/Brase *Understanding Basic Statistics,* Sixth Edition, Student Notetaking Guide

A c ___________________________ for μ is an interval such that c is the percentage of all intervals generated by the same process that contain μ.

Example 1. Find the critical value z_c for $c = 0.90$.

Example 2. Suppose that the standard deviation of all high school seniors' SAT scores in a certain year was σ = 150. A random sample of 100 scores yielded the sample mean $\bar{x} = 1010$. Let μ be the mean of all SAT scores in that year. Find a 0.99 confidence interval for μ. Round your answers to integers.

Example 3. Continue with Example 2.

   (a) What is the value of $z_{0.95}$?

   (b) Is the $\bar{x}$ distribution approximately normal?

   (c) What is the value E when $c = 0.95$? Round your answer to an integer.

   (d) What are the endpoints for a 0.95 confidence interval for μ? Round your answers to integers.

   (e) Interpret the confidence interval.

II. Sample Size for Estimating the Mean μ

Example 4. Suppose that the standard deviation of all high school seniors' SAT scores in a certain year was σ = 150. How large a sample should be taken to be 95% confident that the sample mean $\bar{x}$ is within 50 of the true mean score μ?

<table>
<tr><td>Focus Points
how to compute the
sample size to be
used for estimating a
mean μ</td></tr>
</table>

Homework Assignment

Page(s)

Exercises

Section 8.2 Estimating μ When σ Is Unknown

<table>
<tr><td>Course Number</td></tr>
<tr><td>Instructor</td></tr>
<tr><td>Date</td></tr>
</table>

Objective:

In this section you learned about degrees of freedom and Student's *t* distributions. You also learned how to find critical values using degrees of freedom and confidence levels, how to compute confidence intervals for μ when σ in unknown, and how to explain the meaning of that confidence interval.

Important Vocabulary   Define each term or concept.

Student's *t* distribution

Student's *t* variable

Degrees of freedom

Critical values, t_c

I. Student's *t* Distributions

Focus Points
how to determine degrees of freedom, when to use Student's *t* distributions, how to find critical values using degrees of freedom and confidence levels

A ___________________ is used to obtain information from samples of populations with unknown standard deviation. The distribution is ___________ about the mean 0, is bell-shaped (with thicker tails), and depends on the ___________________. As the degrees of freedom ___________, the *t* distribution approaches the standard normal distribution.

The Student's *t* variable is calculated the formula $t =$ ___________ and the formula for the degrees of freedom is $d.f. =$ ___________.

Example 1. Find the critical value t_c for a 0.90 confidence level for a t distribution with sample size $n = 8$. What is the degree of freedom in this case?

Example 2. Find the values of $t_{0.95}$ and $t_{0.98}$ for a sample of size 5.

II. Confidence Interval For μ When σ Is Unknown

Focus Points
how to compute confidence intervals for μ when σ in unknown, how to explain the meaning of that confidence interval

Example 3. Suppose that the standard deviation of all high school seniors' SAT scores in a certain year was unknown. A sample of 10 students' scores shows these scores:

$$900, 1000, 850, 1200, 1030, 940, 600, 1350, 1400, 700$$

Find a 99% confidence interval for μ, the mean score of all SAT scores of that year. Round your answers to integers.

Example 4. Continue with Example 3.

(a) How many degrees of freedom does this sample have? What are the values of $\bar{x}$ and s of this sample?

(b) What is the $t_{0.99}$ used in Example 3?

(c) If we want a 95% confidence interval instead, what is $t_{0.95}$?

(d) Find a 95% confidence interval for μ. What is the value E in this case? Round your answers to integers.

(e) Interpret the confidence interval obtained in part (d) in the context of the problem.

Homework Assignment

Page(s)

Exercises

Section 8.3 Estimating *p* in the Binomial Distribution

Course Number

Instructor

Date

Objective:

In this section you learned how to compute the maximal margin of error for proportions using a given level of confidence, how to compute confidence intervals for *p* and interpret the results, how to interpret poll results, how to compute the sample size to be used for estimating a proportion *p* when we have an estimate for *p*, and how to compute the sample size to be used for estimating a proportion *p* when we have no estimate for *p*.

Important Vocabulary   Define each term or concept.

Point estimate for *p*

Maximal margin of error for estimating *p*

Confidence interval for *p*

Sample Size for estimating *p*

---

1. Confidence Interval for *p*

Focus Points
how to compute the maximal margin of error for proportions using a given level of confidence, how to compute confidence intervals for p and interpret the results

For a binomial distribution, $\hat{p} = $ _________ can be used as a point estimate for *p*, the population proportion of successes.

The maximal margin of error for estimating *p* equals $E = $ _______.

The confidence interval at confidence level *c* for *p* is an interval such that c ____________ of all such intervals generated using the same procedure contain *p*.

Example 1. Suppose we tossed an *unfair* coin 100 times and found that Heads occurred 60 times. Let *p* be the probability that Heads will occur and *q* be the probability that Tails will occur.

(a) What is the number of trials *n*? What is the value of *r*?

                Brase/Brase *Understanding Basic Statistics,* Sixth Edition, Student Notetaking Guide

(b) What are the point estimates for p and q?

(c) Is the number of trials large enough to justify a normal approximation to the binomial?

(d) Find a 95% confidence interval for p.

Example 2. Continue with Example 1.

(a) Find a 90% confidence interval for p.

(b) Interpret the confidence interval you just computed.

II. Interpreting Results from a Poll

Focus Points
how to interpret poll
results

Example 3. A recent poll based on interviews of 300 residents in a town shows that 210 of them support a town tax cut. The town states that chances are 9 of 10 that if every town resident was interviewed, the findings would differ from the poll results by no more than 2 percentage points in either direction.

(a) What confidence level does the town have?

(b) What is the point estimate for the percentage p of the residents that support a tax cut?

(c) What confidence interval for p does the town have?

III. Sample Size for Estimating p

Focus Points
how to compute the sample size to be used for estimating a proportion p

When estimating a proportion p, if you have a preliminary estimate for p, then your sample size should be $n =$ ___________.

When estimating a proportion p, if you *do not* have a preliminary estimate for p, then your sample size should be $n =$ ___________.

Example 4. Continue with Example 1. Suppose we want to be 99% sure that the point estimate r/n for p will be in error either way by less that 0.01.

(a) If no preliminary study is made to estimate p, how large a sample should be used? That is, how many times do we have to toss the coin?

(b) If we use the value of r obtained in Example 1 as a preliminary estimate for p, then how many times do we have to toss the coin?

Additional Notes

Homework Assignment

Page(s)

Exercises

Chapter 9 Hypothesis Testing

Section 9.1 Introduction to Statistical Tests

Course Number

Instructor

Date

Objective:
In this section you learned the rational for statistical tests. You learned how to identify the null and alternate hypotheses in a statistical test, how to identify right-tailed, left tailed, and two-tailed tests, how to use a test statistic to compute a P-value, how to recognize types of errors, level of significance, and power of a test. You also learned the meaning and risks of rejecting or not rejecting the null hypothesis.

Important Vocabulary   Define each term or concept.

Hypotheses

Hypothesis testing

Null hypothesis

Alternate hypothesis

Left-tailed test

Right-tailed test

Two-tailed test

P-value

Level of significance

Power of a test

Statistical significance

I. Stating Hypotheses

Hypotheses are assertions that you assume to be _____ for the purposes of investigation. Hypothesis testing is used to examine the _________ of a hypothesis. The main question in hypothesis testing is whether it is reasonable to believe that the value of the sample test statistic happens by chance alone to be so _________ from the value of the population parameter proposed in the null hypothesis.

The ___________ is the statement that is under investigation or being tested and is denoted by ___ . The ___________________ is the statement you will adopt in the situation in which the evidence (data) is so strong that you reject the null hypothesis, and is denoted by _____. A statistical _______ is designed to assess the strength of the evidence (data) against the null hypothesis.

> **Focus Points**
> what the rationale is behind statistical tests and how to identify the null and alternate hypotheses in a statistical test

Example 1. A car dealer tells a customer that a certain model of car gets 35 miles per gallon in town, but the customer suspects that the mileage might be overrated. Let μ be the mean of the mileage distribution of this model of cars.

   (a) What should the customer use for H_0?

   (b) What should the customer use for H_1?

Example 2. A company makes doors and windows. The average width of a certain kind of door should be 36 in. To check if the average width is correct, the company formulates a statistical test.

   (a) What should be used for H_0?

   (b) What should be used for H_1?

II. Types of Tests

A statistical test is _________ if H_1 states that the parameter is less than the value claimed in H_0. A statistical test is ____________ if H_1 states that the parameter is greater than the value claimed in H_0. A statistical test is ____________ if H_1 states that the parameter is not equal to the value claimed in H_0.

> **Focus Points**
> how to identify right-tailed, left tailed, and two-tailed tests

III. Hypothesis Tests of μ, Given x Is Normal and σ Is Known, And the *P*-value of a Statistical Test

Focus Points
How to conduct a z test and how to use a test statistic to compute a *P*-value

Assuming H_0 is true, the probability that the test statistic will take on values as extreme as or more extreme than the observed test statistic (computed from sample data) is called the ________ of the test. The smaller the *P*-value computed from sample data, the stronger the evidence against H_0.

Example 3. Continue with Example 1. Suppose we know that the mileage, x, has a normal distribution with the standard deviation $\sigma = 2$ mpg. Suppose a sample of 9 cars of this model were tested and the average mileage of this sample was $\bar{x} = 32.8$ mpg in town. Based on the sample information, should the customer believe the dealer's statement that the mileage is 35 mpg in town?

   (a) Restate the hypotheses.

   (b) Are the observed sample data compatible with H_0 ?

   (c) Assuming H_0 is true, what is the probability for $\bar{x} \leq 32.8$? That is, what is the *P*-value in this test?

   (d) What should be the customer's conclusion?

   (e) Does the test result *prove* that H_0 is false and H_1 is true?

IV. Types of Errors

Focus Points
how to describe types of errors, level of significance, the power of a test

Rejecting H_0 when it is actually true is called a ________ error. The probability with which we are willing to risk a type I error is called the ____________________ of a test. It is denoted by α.

Accepting H_0 when it is actually false is called a ________ error. The probability with which we are willing to risk a type II error is denoted by β.

The quantity $1 - \beta$ is called the ____________ and represents the probability of rejecting the null hypothesis when it is, in fact, false.

Example 4. Continue with Example 2.

(a) Restate the hypotheses.

(b) Suppose the company requires a 5% level of significance. Describe a type I error, its consequence, and its probability.

(c) Discuss a type II error and its consequences.

V. Concluding a Statistical Test

In statistical work, __________________ means that at the alpha level of risk, the evidence (sample data) against the null hypothesis is sufficient to discredit it, so we adopt the alternate hypothesis. We do not claim that we have ________ or __________ the null hypothesis, only that the probability of a type I error (rejecting the null hypothesis when it is, in fact, true) is alpha.

> **Focus Points**
> The meaning and risks of rejecting or not rejecting the null hypothesis

Example 5. Continue with Example 2. Suppose we know that the width of these doors, x, has a normal distribution with a standard deviation $\sigma = 0.5$ in. Suppose a sample of 10 doors was examined and the mean width $\bar{x}$ of this sample was 36.3 in. The company wants to conduct a statistical test with $\alpha = 0.01$.

(a) Restate the hypotheses.

(b) What type of test is this?

(c) What sampling distribution should the company use? What is the value of the sample test statistic?

(d) What is the P-value in this case?

(e) Compare the P-value with the level of significance α. What is the conclusion of the test?

(f) Interpret the results in the context of this problem.

Additional Notes

<table>
<tr><td>

Homework Assignment

Page(s)

Exercises

</td></tr>
</table>

       Brase/Brase *Understanding Basic Statistics,* Sixth Edition, Student Notetaking Guide

Section 9.2 Testing the Mean μ

<table>
<tr><td>Course Number</td></tr>
<tr><td>Instructor</td></tr>
<tr><td>Date</td></tr>
</table>

Objective:

In this section you reviewed the general procedure for testing using
P-values. You learned how to test μ when σ is known using the normal
distribution, and how to test μ when σ is unknown using a Student's *t*
distribution. You also learned the traditional method of testing that uses
critical regions and critical values instead of *P*-values.

Important Vocabulary   Define each term or concept.

Critical region method

Critical values

I. Testing μ When σ Is Known

Focus Points
how to test μ
when σ is
known using
the normal
distribution

Example 1. A car dealer tells a customer that a certain model of car
gets 30 miles per gallon in town, but the customer suspects that the
mileage might be overrated. Let μ be the mean of the mileage
distribution of this model of cars. Suppose we know that the standard
deviation of the mileage, x, equals σ = 5 mpg. Suppose a sample of
40 cars of this model were tested and the average mileage of this sample
was $\bar{x}$ = 28 mpg in town. Based on the sample information, should the
customer believe the dealer's statement that the mileage is 30 mpg in
town? Use α = 0.01.

(a) Establish the hypotheses.

(b) Compute the test statistic from the sample data.

(c) What is the *P*-value in this test?

(d) Conclude the test.

(e) Interpret the results.

II. Testing μ When σ Is Unknown

Example 2. A car dealer tells a customer that a certain model of car gets 30 miles per gallon in town, but the customer suspects that the mileage might be overrated. Let μ be the mean of the mileage distribution of this model of cars. Suppose the standard deviation of the mileage, x, is unknown. Suppose a sample of 41 cars of this model were tested and the average mileage of this sample was $\bar{x} = 28$ mpg and the sample standard deviation of this sample was $s = 8$. Based on the sample information, should the customer believe the dealer's statement that the mileage is 30 mpg in town? Use $\alpha = 0.01$.

> **Focus Points**
> how to test μ
> when σ is
> unknown using
> a Student's t
> distribution

(a) Establish the hypotheses.

(b) Compute the test statistic from the sample data.

(c) How many degrees of freedom are there in this t test?

(d) Find the P-value or the interval containing the P-value.

(e) Conclude the test.

(f) Interpret the results.

Example 3. A company makes doors and windows. The average width of a certain kind of door should be 36 in. To check if the average width is correct, a sample of 10 doors was examined. The mean width $\bar{x}$ of this sample was 36.3 in and the standard deviation of this sample was $s = 0.5$. Does the company need to adjust the production process to meet the width requirement based on this sample? Use $\alpha = 0.05$.

(a) Establish the hypotheses.

(b) Compute the test statistic from the sample data.

(c) How many degrees of freedom are there in this t test?

(d) Find the P-value or the interval containing the P-value.

(e) Conclude the test.

   Brase/Brase *Understanding Basic Statistics,* Sixth Edition, Student Notetaking Guide

(f) Interpret the results.

III. Testing μ Using Critical Regions (Traditional Method)

The values of a distribution for which we reject the null hypothesis make up the _________________ of the distribution. Critical values are the _______________ of the critical region.

> **Focus Points**
> how to use the traditional method of testing involving critical regions and critical values instead of *P*-values

Example 4. Repeat Example 1 with the critical regions method.

(a) Establish the hypotheses.

(b) Compute the test statistic from the sample data.

(c) Determine the critical region and critical value.

(d) Conclude the test.

(e) Interpret the results.

(f) How do results of the critical region method compare to the results of the *P*-value method in this case?

Additional Notes

<table>
<tr><td>

Homework Assignment

Page(s)

Exercises

</td></tr>
</table>

     Brase/Brase *Understanding Basic Statistics,* Sixth Edition, Student Notetaking Guide

Section 9.3 Testing a Proportion *p*

Objective:
In this section you learned how to identify the components needed for
testing a proportion, how to compute the sample test statistic, and how to
find the *P*-value and conclude the test.

Course Number

Instructor

Date

Important Vocabulary   Define each term or concept.

Criteria for using normal approximation to binomial

I. How to Test a Proportion *p*

When $np >$ ___ and $nq >$ __, the normal distribution can be used to
approximate the binomial distribution in testing a proportion *p*.

Example 1. A new medical procedure has been developed to treat a
certain disease. Under the old procedure, it is known that 40% of the
patients get cured. Suppose the new procedure has been used to treat
a sample of 200 patients and 90 of them are cured. Can we justify
that the new procedure is better than the old one? Use a 5% level of
significance.

Focus Points
how to identify
the components
needed for
testing a
proportion, how
to compute the
sample test
statistic, and how
to find the
P-value and
conclude the test

(a) Establish the hypotheses.

(b) Compute the test statistic from the sample data.

(c) What is the *P*-value in this test?

(d) Conclude the test.

(e) Interpret the results.

Example 2. Continue with Example 1. Suppose the doctors keep trying out the new
procedure. After a month, the new procedure has been used to treat 500 more patients and
225 of them are cured. Now can we justify that the new procedure is better than the old one?
Once again, use a 5% level of significance.

(a) Establish the hypotheses.

(b) Compute the test statistic from the sample data.

(c) Find the P-value in this test.

(d) Conclude the test.

(e) Interpret the results.

Example 3. Repeat Example 2 with the critical regions method.

(a) Establish the hypotheses.

(b) Compute the test statistic from the sample data.

(c) Determine the critical region and critical value.

(d) Conclude the test.

(f) Interpret the result.

(e) How do the results of the critical region method compare to the results of the P-value method in this case?

<table>
<tr><td>Homework Assignment</td></tr>
<tr><td>Page(s)</td></tr>
<tr><td>Exercises</td></tr>
</table>

 Brase/Brase *Understanding Basic Statistics,* Sixth Edition, Student Notetaking Guide

Chapter 10 Inferences About Differences

Section 10.1 Tests Involving Paired Differences (Dependent Samples)

<table>
<tr><td>Course Number</td></tr>
<tr><td>Instructor</td></tr>
<tr><td>Date</td></tr>
</table>

Objective:
In this section you learned how to identify paired data and dependent samples, how to explain the advantage of paired data tests, how to compute differences and the sample test statistic, and how to estimate the P-value and conclude the test.

Important Vocabulary   Define each term or concept.

Paired data

Dependent samples

I. Test Paired Differences Using The *t* Distribution

__________ data can be used when there is a natural matching of characteristics. Advantages of using paired data include ______ the danger of introducing extraneous or uncontrollable factors into sample measurements. Paired data have essentially the same characteristics except for the one characteristic that is being measured.

Focus Points
How to identify paired data and dependent samples, how to explain the advantage of paired data tests, how to compute differences and the sample test statistic, and how to estimate the P-value and conclude the test

Example 1. A test preparation school claims that its SAT preparation class will help students improve their SAT scores. To estimate the effectiveness of the class, the school has students take a practice SAT test at the beginning of the class and then another one at the end of the class. A sample of 10 students is taken and their scores of "before" and "after" are compared. Is there a natural way of pairing the scores? How many pairs will the school have?

Example 2. To uncover the natural pattern of temperature change, student is asked to measure the outdoor temperature twice a day, once at 6:00 AM and once at 12:00 noon, for 30 days. The student is then supposed to compare the "morning temperature" with the "noon temperature" based on this sample. Is there a natural way of pairing the temperature measurements? How many pairs will the student have?

Example 3. Continue with Example 1. Below are the "before" and "after" scores of those 10 students, as well as the difference between the two scores.

Before	After	Difference (After score – Before score)
800	1000	200
950	1010	60
1400	1400	0
1200	1250	50
800	900	100
1350	1300	−50
1300	1400	100
900	1050	150
1100	1150	50
1400	1500	100

Based on this sample, conduct a t test to see whether the class helps to improve SAT scores. Use $\alpha = 0.01$

(a) Establish the hypotheses.

(b) Find the value of $\bar{d}$ and s_d.

(c) Find the test statistic t.

(d) How many degrees of freedom are there in this case?

(e) Find the P-value or an interval containing the P-value.

(f) Conclude the test.

(g) Interpret the results.

Brase/Brase *Understanding Basic Statistics,* Sixth Edition, Student Notetaking Guide

Example 4. Continue with Example 3. Suppose the "before" and "after" scores of those 10 students, as well as the difference between the two scores, were in fact as shown below.

Before	After	Difference (After score – Before score)
800	980	180
950	990	40
1400	1380	-20
1200	1230	30
800	880	80
1350	1280	-70
1300	1380	80
900	1030	130
1100	1130	30
1400	1480	80

Based on this sample, conduct a t test to see whether the class helps to improve SAT scores. Use $\alpha = 0.01$

(a) Establish the hypotheses.

(b) Find the value of $\bar{d}$ and s_d.

(c) Find the test statistic t.

(d) How many degrees of freedom are there in this case?

(e) Find the P-value or an interval containing the P-value.

(f) Conclude the test.

(g) Interpret the results.

Example 5. Repeat Example 4 using the critical region method.

(a) Restate the hypotheses.

(b) Find the value of $\bar{d}$ and s_d.

(c) Find the test statistic t.

(d) How many degrees of freedom are there in this case?

(e) Find the critical value for $\alpha = 0.01$ and $d.f. = 9$.

(f) Conclude the test.

(g) Interpret the results.

(h) Is the conclusion here the same as the conclusion from the
P-value method?

Additional Notes

Homework Assignment

Page(s)

Exercises

Brase/Brase *Understanding Basic Statistics,* Sixth Edition, Student Notetaking Guide

Section 10.2 Inferences About the Difference of Two Means $\mu_1 - \mu_2$

<table>
<tr><td>Course Number</td></tr>
<tr><td>Instructor</td></tr>
<tr><td>Date</td></tr>
</table>

Objective:

In this section you learned how to identify independent samples and sampling distributions, how to compute the sample test statistic and *P*-value for testing $\mu_1 - \mu_2$, and how to find confidence intervals for $\mu_1 - \mu_2$.

Important Vocabulary　　　　　Define each term or concept.

Independent samples

Pooled standard deviation

I. Independent Samples

If there is a relationship between corresponding data values in the two distributions, then these two sampling distributions are _____________ . If sample data drawn from one population are completely unrelated to the selection of sample data from the other population, then the two samples are _________________ .

Focus Points
how to identify
independent samples

Example 1. For each experiment below, determine whether the samplings are dependent or independent.

(a) Two different tests are given to two different group of students. Are the test scores from these two groups dependent or independent?

(b) Before a course starts, students are given a pretest. After the course finishes, the same students are given a test covering similar materials. Are the test scores from these two tests dependent or independent?

II. Hypothesis Tests and Confidence Intervals for $\mu_1 - \mu_2$ (σ_1 and σ_2 known)

Example 2. Two competing models of cars are to be compared on the life time of their engine. The records

Focus Points
how to compute the sample test statistic and *P*-value for testing $\mu_1 - \mu_2$, and how to find confidence intervals for $\mu_1 - \mu_2$ (when standard deviations are known)

of 15 cars of the first model and 20 cars of the second model are examined and the following results are found:

Model 1: mean time $\bar{x}_1 = 12$ yr, $\sigma_1 = 2$ yr; $n_1 = 15$

Model 2: mean time $\bar{x}_2 = 11$ yr, $\sigma_2 = 3$ yr; $n_2 = 20$

Assume that the life time of engine is normally distributed. Test whether there is any difference (either way) in the life time of engine between these two models. Use a 1% level of confidence.

(a) Establish the hypotheses.

(b) Find the test statistic z.

(c) Find the P-value.

(d) Conclude the test.

(e) Interpret the results.

Example 3. Continue with Example 2. Find a 99% confidence interval for the difference $\mu_1 - \mu_2$.

Example 4. Interpret a confidence interval.

(a) Suppose a study reported a 95% confidence interval for $\mu_1 - \mu_2$ to be $5 < \mu_1 - \mu_2 < 8$. What can you conclude about $\mu_1 - \mu_2$?

(b) Suppose a study reported a 99% confidence interval for $\mu_1 - \mu_2$ to be $-5 < \mu_1 - \mu_2 < -1$. What can you conclude about $\mu_1 - \mu_2$?

III. Hypothesis Tests and Confidence Intervals for $\mu_1 - \mu_2$ (σ_1 and σ_2 unknown)

Example 5. Two competing models of cars are to be compared on the life time of their engine. The records of 15 cars of the first model and 20 cars of the second model are examined and the following results are found:

Model 1: mean time $\bar{x}_1 = 13.4$ yr, $s_1 = 2$ yr; $n_1 = 15$

Model 2: mean time $\bar{x}_2 = 11$ yr, $s_2 = 3$ yr; $n_2 = 20$

Assume that the life time of engine is normally distributed. Test whether there is any difference (either way) in the life

> **Focus Points**
> how to compute the sample test statistic and P-value for testing $\mu_1 - \mu_2$, and how to find confidence intervals for $\mu_1 - \mu_2$ (when standard deviations are unknown)

Brase/Brase *Understanding Basic Statistics,* Sixth Edition, Student Notetaking Guide

time of engine between these two models. Use a 1% level of confidence.

    (a) Establish the hypotheses.

    (b) Find the test statistic t.

    (c) What is the degree of freedom in this case?

    (d) Find the P-value or an interval containing the P-value.

    (e) Conclude the test.

    (f) Interpret the results.

Example 6. Continue with Example 5. Find a 99% confidence interval for the difference $\mu_1 - \mu_2$.

Example 7. Continue with Example 5. This time test whether μ_1 is greater than μ_2. Still use a 1% level of confidence.

    (a) Establish the hypotheses.

    (b) Find the test statistic t.

    (c) What is the degree of freedom in this case?

    (d) Find the P-value or an interval containing the P-value.

    (e) Conclude the test.

    (f) Interpret the results.

Example 8. Continue with Example 7. Find a 99% confidence interval for the difference $\mu_1 - \mu_2$.

IV. Testing $\mu_1 - \mu_2$ Using Critical Regions (Optional)

Example 9. Repeat Example 5 using the method of critical regions.

> *Focus Points*
> how to test $\mu_1 - \mu_2$ using critical region method

(a) Establish the hypotheses.

(b) Find the test statistic t.

(c) What is the degree of freedom in this case?

(d) What is the t critical value?

(e) Conclude the test.

(f) Interpret the results.

Additional Notes

Homework Assignment

Page(s)

Exercises

 Brase/Brase *Understanding Basic Statistics,* Sixth Edition, Student Notetaking Guide

Section 10.3 Inferences About the Difference of Two Proportions $p_1 - p_2$

<table>
<tr><td>Course Number</td></tr>
<tr><td>Instructor</td></tr>
<tr><td>Date</td></tr>
</table>

Objective:
In this section you learned how to how to compute the sample test statistic and P-value for testing $p_1 - p_2$, and how to find confidence intervals for $p_1 - p_2$.

Important Vocabulary   Define each term or concept.

Pooled estimate of proportion

I. Testing a Difference of Proportions $p_1 - p_2$

Focus Points
how to compute the sample test statistic and P-value for testing $p_1 - p_2$

If two distributions are assumed to have ___________ proportions of successes, you can use a pooled best estimate, which is the sum of the observed number of successes for both trials divided by the sum of the total combined number of trials.

Example 1. The lifetimes of the engines of two competing models of cars are to be compared. The proportion of the engines that last more than 10 years is used for the comparison. The records of 150 cars of the first model and 200 cars of the second model are examined and the following results are found:

    Model 1: Out of the 150 car engines, 90 lasted more than
            10 years;
    Model 2: Out of the 200 car engines, 100 lasted more than
            10 years.

Use a 5% level of significance to test whether the proportion of engines that last more than 10 years of the first model is greater than that of the second model.

(a) Establish the hypotheses.

(b) Find the sample statistic $\hat{p}_1 - \hat{p}_2$.

(c) Find the test statistic z.

(d) Find the P-value.

(e) Conclude the test.

Brase/Brase *Understanding Basic Statistics*, Sixth Edition, Student Notetaking Guide   **105**

(f) Interpret the results.

(g) If the critical region method is used, what is the critical value
in this test? What will be the conclusion of the test?

Example 2. Repeat Example 1 with $\alpha = 1\%$.

(a) Establish the hypotheses.

(b) Find the sample statistic $\hat{p}_1 - \hat{p}_2$.

(c) Find the test statistic z.

(d) Find the P-value.

(e) Conclude the test.

(f) Interpret the results.

(g) If the critical region method is used, what is the critical value
in this test? What will be the conclusion of the test?

II. Estimating a Difference of Proportions $p_1 - p_2$

Example 3. Continue with Example 1. Find a 95% confidence
interval for $p_1 - p_2$.

> *Focus Points*
> how to find confidence
> intervals for $p_1 - p_2$.

Homework Assignment

Page(s)

Exercises

Brase/Brase *Understanding Basic Statistics,* Sixth Edition, Student Notetaking Guide

Chapter 11 Additional Topics Using Inference

Section 11.1 Chi-Square: Tests of Independence and of Homogeneity

<table>
<tr><td>Course Number</td></tr>
<tr><td>Instructor</td></tr>
<tr><td>Date</td></tr>
</table>

Objective:
In this section you learned how to set up a test to investigate independence of random variables, how to use contingency tables to compute the sample χ^2 statistic, how to find or estimate the P-value of the sample χ^2 statistic and complete the test, and how to conduct a test of homogeneity of populations.

Important Vocabulary   Define each term or concept.

Chi-square distribution

Contingency table

Expected frequency

Test of homogeneity

I. Test of Independence

The chi-square distribution is __________________ and varies depending on the __________________.

When comparing two factors, a ________________ table is used to record the ____________________, which can be pre-calculated for each cell in the table by making assumptions about the probabilities for each factor.

> **Focus Points**
> how to set up a test to investigate independence of random variables, how to use contingency tables to compute the sample χ^2 statistic, how to find or estimate the P-value of the sample χ^2 statistic and complete the test

Under the null hypothesis that two factors are independent, we can calculate the expected frequency of a cell by _______________ together the probabilities of each factor which is assumed to be independent.

Example 1. Suppose a contingency table has four rows and three columns, not counting the row and column totals. What is the size of this contingency table? How many cells are there in the table?

Example 2. A college wants to see if there is any relationship between students' performance on tests and the time of the day when the tests are given. A course with multiple sections was chosen and students in all sections were given the same test. Some sections were tested in the morning and some in the afternoon. The scores were grouped into three categories: A-or-B, C, D-or-below. The sample results are listed in the table below. Find the expected frequency for cell #2.

Time vs. Scores	A-or-B	C	D-or-below	Row Total
Morning	#1 50	#2 40	#3 30	120
Afternoon	#4 30	#5 20	#6 30	80
Column Total	80	60	60	200

Example 3. Continue with Example 2. Complete the contingency table.

Time vs. Scores	A-or-B	C	D-or-below	Row Total
Morning	#1 $O = 50$ $E = 48$	#2 $O = 40$ $E =$ ___	#3 $O = 30$ $E =$ ___	120
Afternoon	#4 $O = 30$ $E =$ ___	#5 $O = 20$ $E =$ ___	#6 $O = 30$ $E =$ ___	80
Column Total	80	60	60	200

Example 4. Continue with Example 3. Complete the following table of differences between observed values and expected frequencies.

Cell	O	E	$O - E$	$(O - E)^2$	$(O - E)^2 / E$
1	50	48	2	4	___
2	40	___	___	___	___
3	30	___	___	___	___
4	30	___	___	___	___
5	20	___	___	___	___
6	30	___	___	___	___
				$\sum (O - E)^2 / E =$	___

     Brase/Brase *Understanding Basic Statistics,* Sixth Edition, Student Notetaking Guide

Example 5. Continue with Example 4. Compute the test statistic χ^2 for this sample.

Example 6. Continue with Example 5. How many degrees of freedom are there in this example?

Example 7. Continue with Example 6. Test whether students' performance on tests and the time of the day when the tests are given are independent. Use a level of significance of $\alpha = 0.05$.

(a) State the hypotheses.

(b) Estimate the *P*-value.

(c) Conclude the test.

(d) Interpret the test.

II. Tests of Homogeneity

<table>
<tr><td>

A test of ________________ tests the claim that different populations share the same ________________ of specified characteristics.

</td><td>

Focus Points
how to conduct a test of homogeneity of populations

</td></tr>
</table>

Example 8. Continue with Example 7. Test whether the proportions of score groups in the morning tests and those in the afternoon tests are the same. Use a level of significance at $\alpha = 0.05$.

(a) State the hypotheses.

(b) Estimate the *P*-value.

(c) Conclude the test.

(d) Interpret the test.

Additional Notes

Homework Assignment

Page(s)

Exercises

Section 11.2 Chi-Square: Goodness of Fit

<table>
<tr><td>Course Number</td></tr>
<tr><td>Instructor</td></tr>
<tr><td>Date</td></tr>
</table>

Objective:

In this section you learned how to set up a test to investigate how well a sample distribution fits a given distribution, how to use observed and expected frequencies to compute the sample χ^2 statistic, and how to find or estimate the P-value and complete the test.

Important Vocabulary   Define each term or concept.

Goodness-of-fit test

I. Test for Goodness of Fit

The ______________________________ is used to determine whether a population fits a given distribution. For goodness of fit tests, we use a ______________ test on the chi-square distribution.

Focus Points
how to set up a test to investigate how well a sample distribution fits a given distribution, how to use observed and expected frequencies to compute the sample χ^2 statistic, how to find or estimate the P-value and complete the test

Example 1. A college president thinks that the most appropriate student grade distribution should be A – 15%; B – 20%; C –30%; D – 20%; and F – 15%. He takes a random sample of 100 students from a certain large class, and looks at the course grades of these students. He finds that of these 100 students, 20 got an A in the class, 25 got a B, 25 got a C, 20 got a D, and 10 got an F. Test whether the population of student grades follows the distribution that the president would like to see. Use a 5% level of significance.

(a) What is the distribution that the president would like to see?

	A	B	C	D	F
Probability *P*					

(b) State the test hypotheses:

(c) Complete the following table and use the table to compute the sample statistic χ^2.

Score	O	E	$O - E$	$(O - E)^2$	$(O - E)^2/E$
A	20	15	5	25	___
B	25	___	___	___	___
C	25	___	___	___	___
D	20	___	___	___	___
F	10	___	___	___	___
			$\chi^2 = \sum (O - E)^2 / E =$ ___		

(d) How many degrees of freedom are there in this case?

(e) Find or estimate the P-value.

(f) Conclude the test.

(g) Interpret the test.

Additional Notes

Homework Assignment

Page(s)

Exercises

    Brase/Brase *Understanding Basic Statistics,* Sixth Edition, Student Notetaking Guide

Section 11.3 Testing a Single Variance or Standard Deviation

Course Number

Instructor

Date

Objective:

In this section you learned how to set up a test for a single variance σ^2, how to compute the sample χ^2 statistic, and how to use the χ^2 distribution to estimate the P-value and conclude the test.

Important Vocabulary   Define each term or concept.

I. Testing σ^2

Focus Points
how to set up a test for a single variance σ^2, how to compute the sample χ^2 statistic, and how to use the χ^2 distribution to estimate the P-value and conclude the test

Example 1. A company makes a special kind of windows. Historically, the standard deviation of the width of the kind of window made by this company is 0.4 inches. The company recently started a new production procedure. A random sample of 30 windows made with the new procedure was examined, and it was found that the standard deviation of the width of windows in this sample was only 0.2 inches. Use $\alpha = 0.01$ to test the claim that the variance in width is reduced by using the new procedure.

(a) State the hypotheses:

(b) Find the value of χ^2 and the degrees of freedom.

(c) Estimate the P-value.

(d) Conclude the test.

(e) Interpret the conclusion.

Example 2. Let x be the height in inches of 16-year-old boys in a certain country. Ten years ago, the population standard deviation of x was approximately 0.4 inches. Suppose that a recent interview of 30 boys of 16 years old in that country gave a sample standard deviation of 0.33 inches. Test whether the current variance is different from that of ten years ago. Use a 1% level of significance.

(a) State the hypotheses:

(b) Find the value of χ^2 and the degrees of freedom.

(c) Estimate the P-value.

(d) Conclude the test.

(e) Interpret the conclusion.

Example 3. A new way of treating a certain kind of cancer is compared with the old one. The old method was known to keep the patient alive for a mean of 10 years with a standard deviation of 4 years. To test the variability of the new method, a random sample of 30 patients given the new treatment was studied and the results showed a sample standard deviation of 4.5 years. Use $\alpha = 0.01$ to test whether the variance has increased under the new treatment.

(a) State the hypotheses:

(b) Find the value of χ^2 and the degrees of freedom.

(c) Estimate the P-value.

(d) Conclude the test.

(e) Interpret the conclusion.

<table>
<tr><td>Homework Assignment</td></tr>
<tr><td>Page(s)</td></tr>
<tr><td>Exercises</td></tr>
</table>

Section 11.4 Inferences for Correlation and Regression

<table>
<tr><td>Course Number</td></tr>
<tr><td>Instructor</td></tr>
<tr><td>Date</td></tr>
</table>

Objective:
In this section you learned how to test the correlation coefficient ρ, how to use sample data to compute the standard error of estimate S_e, how to test the slope β of the least-squares line, and how to find a confidence interval for the value of y predicted for a specific value of x.

Important Vocabulary Define each term or concept.

Population correlation coefficient ρ

Standard error of estimate S_e

Confidence prediction band

I. Testing the Correlation Coefficient

Focus Points
how to test the correlation coefficient ρ

The ______________________ can be computed if the (x, y) values are representative of all possible (x, y) pairs, and both x and y values are normally distributed for their paired y and x values.

Example 1. Suppose a sample of thirty (x, y) pairs yields a sample correlation coefficient $r = 0.2$. Test whether the population correlation coefficient ρ is positive. Use a 5% level of significance.

(a) State the hypotheses.

(b) The test statistic $t =$ ______

(c) The degree of freedom = ____

(d) Estimate the P-value.

(e) Conclude the test.

(f) Interpret the conclusion.

II. Standard Error of Estimate

In a scatter diagram, the ___________________, denoted by S_e, , is a common tool for measuring the spread. Let $\hat{y} = a + bx$ be the least-squares line to fit a sample of (x, y) pairs with sample size $n \geq 3$, then the formula for computing S_e is ___________________.

Focus Points
how to use sample data to compute the standard error of estimate S_e

Example 2. For the sample

x	1	2	3	4	5
y	8	9	7	5	4

Find the standard error of estimate S_e.

III. Inferences About the Slope β

Let $\hat{y} = a + bx$ be the sample-based least-squares line. Let $y = \alpha + \beta x$ be the population least-squares line. Let's look at the inference about the slope β.

Focus Points
how to test the slope β of the least-squares line

Example 3. Given the following sample, test whether β is negative. Use a 5% level of significance.

x	1	2	3	4	5
y	8	9	7	6	4

(a) Find the equation of the sample-based least-squares line.

(b) State the hypotheses:

(c) Find the test statistic t and the degrees of freedom.

(d) Estimate the P-value.

(e) Conclude the test.

(f) Interpret the conclusion.

Example 4. Given the sample:

x	10	20	30	40	50
y	21	23	25	23	27

(a) Find the values of the sample correlation coefficient r and the slope b of the sample-based least-squares line.

(b) Use a 1% level of significance to test the claim that $\rho > 0$.

(c) Use a 1% level of significance to test the claim that $\beta > 0$.

IV. Confidence Intervals for y

Example 5. For the sample:

x	10	20	30	40	50
y	21	23	25	23	27

> *Focus Points*
> how to find a
> confidence interval
> for the value of y
> predicted for a
> specific value of x

(a) Find the equation of the sample-based least-squares line.

(b) Find a 90% confidence interval for the value of y when $x = 25$.

Example 6. Continue with Example 5.

(a) What is $\hat{y}$, the predicted value of y, when $x = 35$?

(b) Find the value of the bound E on the error of the estimate in part (a) with a 90% confidence.

(c) Find a 90% confidence interval for the value of y when $x = 35$.

Additional Notes

Homework Assignment

Page(s)

Exercises

Made in the USA
Monee, IL
07 July 2026

56552406R00070